AF614773

ARSENIC - ANALYTICAL AND TOXICOLOGICAL STUDIES

Edited by **Margarita Stoytcheva**
and **Roumen Zlatev**

Arsenic - Analytical and Toxicological Studies
http://dx.doi.org/10.5772/intechopen.72058
Edited by Margarita Stoytcheva and Roumen Zlatev

Contributors

Rebeca Monroy-Torres, Nonhlangabezo Mabuba, Ljubinka Rajakovic, Vladana Rajakovic-Ognjanovic, Jayasri Das Sarma, Afaq Hussain, Vineeth A R, Soumya Kundu, Tapendu Samanta, Raja Shunmugam, Debnath Pal, Margarita Stoytcheva

© The Editor(s) and the Author(s) 2018
The rights of the editor(s) and the author(s) have been asserted in accordance with the Copyright, Designs and Patents Act 1988. All rights to the book as a whole are reserved by INTECHOPEN LIMITED. The book as a whole (compilation) cannot be reproduced, distributed or used for commercial or non-commercial purposes without INTECHOPEN LIMITED's written permission. Enquiries concerning the use of the book should be directed to INTECHOPEN LIMITED rights and permissions department (permissions@intechopen.com).
Violations are liable to prosecution under the governing Copyright Law.

Individual chapters of this publication are distributed under the terms of the Creative Commons Attribution 3.0 Unported License which permits commercial use, distribution and reproduction of the individual chapters, provided the original author(s) and source publication are appropriately acknowledged. If so indicated, certain images may not be included under the Creative Commons license. In such cases users will need to obtain permission from the license holder to reproduce the material. More details and guidelines concerning content reuse and adaptation can be found at http://www.intechopen.com/copyright-policy.html.

Notice

Statements and opinions expressed in the chapters are these of the individual contributors and not necessarily those of the editors or publisher. No responsibility is accepted for the accuracy of information contained in the published chapters. The publisher assumes no responsibility for any damage or injury to persons or property arising out of the use of any materials, instructions, methods or ideas contained in the book.

First published in London, United Kingdom, 2018 by IntechOpen
IntechOpen is the global imprint of INTECHOPEN LIMITED, registered in England and Wales, registration number: 11086078, The Shard, 25th floor, 32 London Bridge Street
London, SE19SG – United Kingdom
Printed in Croatia

British Library Cataloguing-in-Publication Data
A catalogue record for this book is available from the British Library

Additional hard copies can be obtained from orders@intechopen.com

Arsenic - Analytical and Toxicological Studies, Edited by Margarita Stoytcheva and Roumen Zlatev
p. cm.
Print ISBN 978-1-78923-516-6
Online ISBN 978-1-78923-517-3

We are IntechOpen, the world's leading publisher of Open Access books

Built by scientists, for scientists

3,650+
Open access books available

114,000+
International authors and editors

118M+
Downloads

151
Countries delivered to

Our authors are among the
Top 1%
most cited scientists

12.2%
Contributors from top 500 universities

WEB OF SCIENCE™

Selection of our books indexed in the Book Citation Index in Web of Science™ Core Collection (BKCI)

Interested in publishing with us?
Contact book.department@intechopen.com

Numbers displayed above are based on latest data collected.
For more information visit www.intechopen.com

For EU product safety concerns:
IN TECH d.o.o., Prolaz Marije Krucifikse Kozulić 3, 51000 Rijeka, Croatia,
info@intechopen.com or visit our website at intechopen.com.

Meet the editors

Prof. Margarita Stoytcheva graduated from the University of Chemical Technology and Metallurgy in Sofia, Bulgaria, with the title Chemical Engineer and Master of Electrochemical Technologies. She has PhD and DSc degrees in Chemistry and Technical Sciences. She has been actively involved in research and teaching at several universities in Bulgaria, Algeria and France. From 2006 to the present, she has participated in activities of scientific research, technological development and teaching in Mexico at the University of Baja California and the Institute of Engineering, Mexicali. Since 2008, she has been a member of the National System of Researchers in Mexico, and since 2011, she has been a regular member of the Mexican Academy of Sciences. Her interests and areas of research are analytical chemistry and biotechnology.

Dr. Roumen Zlatev is a full-time researcher at the Engineering Institute of the Autonomous University of Baja California (UABC) at Mexicali, Mexico. He obtained his Bachelor's and Master's degrees from the University of Chemical Technology and Metallurgy of Sofia, Bulgaria, and his PhD degree from the National Polytechnic University of Grenoble, France. He was a full-time researcher at the Bulgarian Academy of Sciences and a part-time professor at Sofia University. In 2005, he accepted the position of full-time senior researcher at UABC. Dr. Zlatev is a member of the Mexican National System of Researchers and a regular member of the Mexican Academy of Sciences. He participates in research projects in France, Germany, and Mexico. He is the author of more than 170 publications, book chapters, and reports in scientific congresses, and he holds 14 patents in the field of the electrochemical and spectroscopic methods of analysis, corrosion and materials, electrochemical and analytical instrumentation.

Contents

Preface

Arsenic - Analytical and Toxicological Studies is a compilation of five chapters focused on arsenic occurrence and chemistry, methods for arsenic determination and removal, and arsenic toxicity and health risks.

The three chapters included in the first section introduce: (1) the most commonly used techniques for arsenic quantification and speciation, such as atomic absorption spectrometry, inductively coupled plasma-mass spectrometry, and inductively coupled plasma-emission spectrometry, among others, often conjugated with separation techniques such as high-performance liquid chromatography, and (2) the various procedures applied for arsenic separation and removal, including ion exchange, sorption, coagulation and membrane technologies. The advantages and drawbacks of the methods for arsenic determination and removal are exhaustively described.

The two chapters contained in the second section provide important information about: (1) the mechanism of arsenic-induced toxicology, involving arsenic interaction with critical thiols in proteins, and (2) the health risks associated with exposure to arsenic. The two case studies presented (Guanajuato, Mexico, and Bihar, India) comment in detail on the adverse effects of exposure to arsenic in water.

The book offers a professional multifaceted vision of several problems related to arsenic determination and effects of arsenic exposure.

All the contributing authors are gratefully acknowledged for their time and effort.

Prof. DSc. Eng. Margarita Stoytcheva and Prof. Dr. Eng. Roumen Zlatev
Universidad Autónoma de Baja California
Mexicali, México

Section 1

Arsenic Occurrence and Removal

Chapter 1

Introductory Chapter: Arsenic

Margarita Stoytcheva and Roumen Zlatev

Additional information is available at the end of the chapter

http://dx.doi.org/10.5772/intechopen.78399

1. Arsenic occurrence

Arsenic (As) is a chemical element the abundance of which in the continental crust of the Earth is given as 1.5–2 ppm. Most often arsenic appears in nature in the form of Fe, Co and Ni arsenides, arsenic sulfide, and native arsenic. It belongs to the metalloids, in spite that it shows intermediate properties between metals and non-metals.

In general, arsenic forms soluble oxyanions with the predominant oxidation states +3 and +5, while the minor oxidation states are 0 and −3. The main form of As(III) is arsenite existing in solutions with pH between 5 and 9. H_3AsO_3 and the anions resulting from its dissociation occurring according to Eq. 1 are $H_2AsO_3^{-}$, $H_2AsO_3^{2-}$, and AsO_3^{3-}:

$$H_3AsO_3 \xrightarrow{pK_a=9.2} H_2AsO_3^- + H^+ \xrightarrow{pK_a=14.22} HAsO_3^{2-} + H^+ \xrightarrow{pK_a=19.22} AsO_3^{3-} + H^+ \quad (1)$$

The main form of As(V) is H_3AsO_4. The anions resulting from its dissociation are $H_2AsO_4^{-}$, $HAsO_4^{2-}$, and AsO_4^{3-}. The arsenates have high-ionization capacity due to the presence of the double bond. The H_3AsO_4 molecule easily loses hydrogen ions by dissociation remaining with a negative charge, forming successively several anions according to Eq. 2:

$$\mathrm{H_3AsO_4 \xrightarrow{pK_a=2.2} H_2AsO_4^- + H^+ \xrightarrow{pK_a=6.94} HAsO_4^{2-} + H^+ \xrightarrow{pK_a=11.3} AsO_4^{3-} + H^+} \quad (2)$$

2. Arsenic and its compound application and removal

Elemental arsenic has a limited application mainly as lead and brass alloy's ingredient, but its compounds are widely used in the industry. The gallium arsenide (GaAs) is extensively used in the electronic industry for high-frequency integrated circuits, laser diodes, and Light Emission Diodes (LEDs); the arsenic oxide is used in glass production industry; the arsenic sulfides

© 2018 The Author(s). Licensee IntechOpen. This chapter is distributed under the terms of the Creative Commons Attribution License (http://creativecommons.org/licenses/by/3.0), which permits unrestricted use, distribution, and reproduction in any medium, provided the original work is properly cited.

are used as pigments in textile. There is a revival of the interest in arsenic as medicine for the treatment of acute promyelocytic leukemia, multiple myeloma, myelodysplastic syndrome, and various resistant solid tumors. Other applications of the arsenic compounds include the paper industry, the pyrotechnic industry, and so on. Arsenic compounds were used as pesticides in the past.

The conventional methods most applied for arsenic removal are:

- Coagulation and flocculation (the most commonly employed), which include two distinct processes: colloids destabilization neutralizing the electric charges allowing particles aggregation, followed by flocculation by a polymer building giant aggregates [1, 2].
- Precipitation including one of the following methods: alum coagulation [3], iron coagulation [4], lime softening [5], and combination of iron (and manganese) removal [6–8].
- Membrane filtration: a pressure driven process, which can include one of the following: microfiltration, ultrafiltration, nanofiltration, and hyperfiltration or reverse osmosis [9–12].
- Adsorption and ion exchange are employing solids to accumulate on their surface substances contained in liquid phase [12–16].
- Bioremediation, ozone oxidation, and electrochemical treatments are also applied.

3. Arsenic and its compounds toxicity

Arsenic is extremely poisonous for multicellular life, although several types of bacteria are capable of using arsenic compounds as respiratory metabolites. Pollution of groundwater with arsenic is a problem that affects millions of people around the world. Worldwide, up to 100 million people are at risk of exposure to arsenic from excess arsenic in drinking water in countries such as Argentina, Bangladesh, Chile, China, India, Mexico, and the USA. The US Environmental Protection Agency classifies arsenic as a carcinogen of group A due to the evidence of its adverse effects on health [17]. Exposure to 0.05 mg L^{-1} can cause 31.33 cases of skin cancer per 1000 inhabitants. For this reason, the maximum acceptance limit of 0.050 mg L^{-1} of arsenic was lowered to 0.010–0.020 mg L^{-1}. However, it was found that a daily intake of 12–15 μg arsenic as a microelement from meat, fish, vegetables, and cereals maintains the immune system activity. The WHO warns that the maximum safe arsenic concentration for health is as low as 10 ppb, but limit values for arsenic species are not established [17, 18].

Arsenic can be found as a pollutant in water or food, especially in shells and other seafood, and often polluting fruits and vegetables, especially rice. Today arsenic poisoning occurs through industrial exposure, from contaminated wine or smuggled spirits, or due to intentional use. The possibility of arsenic contamination of herbal preparations and food supplements should also be taken into account.

Inorganic forms of arsenic are more toxic than the organic ones. Arsenic oxides are the most common threat because arsenite and arsenate salts are the most toxic. These forms are components of geological formations and are extracted into groundwater.

The two forms of inorganic arsenic reduced (trivalent arsenic) and oxidized (five valent arsenic) can be absorbed and accumulated in tissues and body fluids. The trivalent form is more toxic and reacts with thiol groups. Very few organs and systems are not affected by the toxic effects of arsenic. The arsenic compound toxicity decays in the following order: As(III) > monomethylarsine oxide > dimethylarson > monomethylarsenate > As(V). The toxicity of As(III) is 10 times greater than that of As(V) and the lethal dose for adults is 1–4 mg kg^{-1}.

4. Analytical methods for arsenic quantification

The "total arsenic" determined in a sample most often do not represent a valuable information, because of the different properties and different toxicity of its species [19, 20]. The main problem is the easy conversion of the arsenic species to another caused by the pH changes, the presence of reducing agents, oxidizers, and certain bacterial strains able to produce a lot of organic As-species [21, 22]. That is why the As speciation is the only important way to characterize the origin of the As-related problems such as its toxicity and biogeochemical cycling and to find the best procedure for drinking water treatment. Unfortunately, As speciation remains challenging, because of the interference between the arsenic species possessing different toxicity [23], which is typical for the organic As-species [24, 25].

The best analytical methods for As speciation are considered those, including chromatographic separations [26, 27] such as IC [28] and HPLC [29], coupled with a sensitive detection system, such as ICP–MS, AFS–HG, and AAS–HG [30]. Specific sorbents and exchange resins have been developed and applied recently for this purpose [31–35]. Apart from the chromatographic and non-chromatographic methods for the arsenic species separation, simple and cost-effective electrochemical methods were developed recently based on the distinct As-species electrochemical properties [36, 37].

5. Conclusion

In spite of the arsenic and its compounds important industrial application, the greatest attention is paid to their influence on the human health, especially in case of long temp action by contaminated drinking water and food. That is why detailed studies on arsenic toxicity and cancer provoking mechanisms were realized. The toxicity of the arsenic species may differ 10 and more times from one to another, so their transformation caused by condition changes (as pH, dissolved oxygen, and bacterial activity) during the sample transportation and conservation prior the laboratory analysis may cause false results. For this reason, simple analytical techniques and sensors for arsenic speciation have been developed able to realize *in-situ* speciation to evaluate the real toxicity of contaminated water, food, or soil samples. To prevent the negative arsenic influence on the human health, the most important is the decontamination of the affected waters and soils. Many methods based on different principles were developed ad successfully applied for this purpose.

Author details

Margarita Stoytcheva* and Roumen Zlatev

*Address all correspondence to: margarita.stoytcheva@uabc.edu.mx

Universidad Autónoma de Baja California, Instituto de Ingeniería, Mexicali, México

References

[1] McNeill LS, Edwards MJ. Soluble arsenic removal at water treatment plants. Journal AWWA. 1995;**87**:105-113

[2] Gregor J. Arsenic removal during conventional aluminium-based drinking-water treatment. Water Research. 2001;**35**:1659-1664

[3] Kartinen EO, Martin CJ. An overview of arsenic removal processes. Desalination. 1995; **103**:79-88

[4] Jones CJ, Hudson BC, McGugan PJ. The removal of arsenic(V) from acidic solutions. Journal of Hazardous Materials. 1977;**2**:333-345

[5] Gaid K. The removal of arsenic from drinking water. Journal Europeen d'Hydrologie. 2005;**36**:145-165

[6] Sorg TJ. Adsorptive media for arsenic removal. In: Public Water System Compliance Using POU and POE Treatment Technology NFS Conference; Orlando, FL; 2003

[7] Sorg TJ, Lytlem D. Arsenic treatment pilot plant tests. Course Materials. In: Water and Wastewater Workshop; Operator Training Committee of Ohio; 2005

[8] Kunzru S, Chaudhuri M. Manganese amended activated alumina for adsorption/oxidation of arsenic. Journal of Environmental Engineering. 2005;**131**:1350-1353

[9] Wiesner MR, Clark MM, Jacangelo JG, Lykins BW, Marinas BJ, O'Melia CR, Rittman BE, Semmens MJ, Brittan J, Fiessinger F, Gemin J, Summers RS, Thompson MA, Tobiason JE. Committee report. Membrane processes in potable water treatment. Journal AWWA. 1992; **84**:59-67

[10] Waypa JJ, Elimelech M, Hering JG. Arsenic removal by RO and NF membranes. Journal AWWA. 1997;**89**:102-114

[11] Seidel A, Waypa JJ, Elimech M. Role of charge (Donnan) exclusion in removal of arsenic from water by a negatively charged porous nanofiltration membrane. Environmental Engineering Science. 2001;**18**:105-113

[12] Saitúa H, Campderrós M, Cerutti S, Pérez A. Effect of operating conditions in removal of arsenic from water by nanofiltration membrane. Desalination. 2005;**172**:173-180

[13] Wennrich RD, Weiss H. Sorption materials for arsenic removal from wastes: a comparative study. Water Research. 2004;**38**:2948-2954

[14] Huang CP, Fu PLK. Treatment of arsenic(V)-containing water by the activated carbon process. Journal of the Water Pollution Control Federation. 1984;**56**:233-242

[15] Gimbel R, Hobby R. Discharge of arsenic and heavy metals from activated carbon filters during drinking water treatment. Wasser Rohrbau. 2000;**51**:15-16

[16] Eguez HE, Cho EH. Adsorption of arsenic on activated-charcoal. Journal of Meteorology. 1987;**39**:38-41

[17] US EPA. National primary drinking water regulations: Arsenic and clarifications to compliance and new source contaminants monitoring. In: Final Rule, Code of Federal Regulations, Title 40, parts 141 and 142, USA. 2001

[18] WHO. Arsenic in Drinking Water. In: Geneva: WHO; 2001

[19] Ammann AA. Arsenic speciation analysis by ion chromatography-a critical review of principles and applications. American Journal of Analytical Chemistry. 2011;**2**:27-45

[20] Smedley PL, Kinniburgh DG. A review of the source, behavior and distribution of arsenic in natural waters. Applied Geochemistry. 2002;**17**:517-568

[21] Feldmann J, Devalla S, Raab A, Hansen HR. In: Hirner AV, Emons H, editors. Organic Metal and Metalloid Species in the Environment. Heidelberg: Springer; 2004. p. 41

[22] Goessler W, Kuehnelt D. In: Mester Z, Sturgeon R, editors. Comprehensive Analytical Chemistry. Amsterdam: Elsevier; 2003. pp. 1027-1044

[23] Miller GP, Norman DI, Frisch PL. A comment on arsenic species separation using ion exchange. Water Research. 2000;**34**:1397-1400

[24] Francesconi KA. Toxic metal species and food regulations-making a healthy choice. Analyst. 2007;**132**:17-20

[25] Nischwitz V, Pergantis SA. Optimisation of an HPLC selected reaction monitoring electrospray tandem mass spectrometry method for the detection of 50 arsenic species. Journal of Analytical Atomic Spectrometry. 2006;**21**:1277-1286

[26] Szpunar J, Lobinski R. Multidimensional approaches in biochemical speciation analysis. Analytical and Bioanalytical Chemistry. 2002;**373**:404-411

[27] Komorowicz I, Baralkiewicz D. Arsenic and its speciation in water samples by high performance liquid chromatography inductively coupled plasma mass spectrometry-last decade review. Talanta. 2011;**84**:247-261

[28] Gault AG, Polya DA, Lythgoe PR, Farquhar ML, Charnock JM, Wogelius RA. Arsenic speciation in surface waters and sediments in a contaminated waterway: An IC-ICP-MS and XAS based study. Applied Geochemistry. 2003;**18**:1387-1397

[29] Ronkart SN, Laurent V, Carbonnelle P, Mabon N, Copin A, Barthélemy JP. Speciation of five arsenic species (arsenite, arsenate, MMAA, DMAA and AsBet) in different kind of water by HPLC-ICP-MS. Chemosphere. 2007;**6**:738-745

[30] Akter KF, Chen Z, Smith L, Davey D, Naidu R. Speciation of arsenic in groundwater samples: a comparative study of CE-UV, HG-AAS and LC-ICP-MS. Talanta. 2005;**68**:406-415

[31] Chen D, Huang C, He M, Separation HB. preconcentration of inorganic arsenic species in natural water samples with 3-(2-aminoethylamino) propyltrimethoxysilane modified ordered mesoporous silica micro-column and their determination by inductively coupled plasma optical emission spectrometry. Journal of Hazardous Materials. 2009;**164**:1146-1151

[32] Issa NB, Rajakovic-Ognjanovic VN, Jovanovic BM, Rajakovic LV. Determination of inorganic arsenic species in natural waters-Benefits of separation and preconcentration on ion exchange and hybrid resins. Analytica Chimica Acta. 2010:185-193

[33] Issa NB, Rajakovic-Ognjanovic VN, Marinkovic AD, Rajakovic LV. Separation and determination of arsenic species in water by selective exchange and hybrid resins. Analytica Chimica Acta. 2011;**706**:191-198

[34] Issa NB, Marinkovic AD, Rajakovic LV. Separation and determination of dimethylarsenate in natural waters. Journal of the Serbian Chemical Society. 2012;**77**:775-788

[35] Daus B, Mattusch J, Wennrich R, Weiss H. Analytical investigations of phenyl arsenicals in groundwater. Talanta. 2008;**75**:376-379

[36] Zlatev R, Stoytcheva M, Valdez B, Álvarez L, Cobo JM. Application of single-use PCB gold disc electrodes for "in-situ" determination of As(III) in natural waters. ECS Transactions. 2009;**20**:3-12

[37] Zlatev R, Stoytcheva M, Valdez B. As(III) determination in the presence of Pb(II) by Differential Alternative Pulses Voltammetry. Electroanalysis. 2010;**22**:1671-1674

Chapter 2

Arsenic in Water: Determination and Removal

Ljubinka Rajakovic and
Vladana Rajakovic-Ognjanovic

Additional information is available at the end of the chapter

http://dx.doi.org/10.5772/intechopen.75531

Abstract

Depending on the physical, chemical and biogeochemical processes and condition of the environment, various arsenic species can be present in water. Water soluble arsenic species existing in natural water are inorganic arsenic (iAs) and organic arsenic (oAs) species. All acidic species, according to the chemical equilibrium, have well-recognized molecular and ionic forms in water. The distribution of iAs and oAs species is a function of pH value of water traces of arsenic that are found in groundwater, lakes, rivers and ocean. The WHO provisional guideline value for arsenic in drinking water is 10 μg L^{-1}. The most selective and sensitive methods for determination of total arsenic and its species in water are coupled techniques including chromatography, optical methods and mass spectrometry. Determination of arsenic species is of crucial importance for selection of arsenic removal technology. Best available technologies are based on absorption, precipitation, membrane and hybrid membrane processes. Adsorption is considered to be relatively simple, efficient and low-cost removal technique, especially convenient for application in rural areas. Sorbents for arsenic removal are biological materials, mineral oxides, activated carbons and polymer resins.

Keywords: arsenic, water, traces, species, toxicity, determination, removal, purification

1. Introduction

Arsenic, As, belongs to the group of elements that are called metalloids. A metalloid is a chemical element that has properties of both metals and nonmetals. Arsenic is from all its features mostly recognized as a poison. Arsenic has a complex chemical behavior since it exists in four different oxidative states [1]. Depending on oxidative state and presence in environment, arsenic species exhibit different toxicity [2]. Arsenic species can be present in all types of

© 2018 The Author(s). Licensee IntechOpen. This chapter is distributed under the terms of the Creative Commons Attribution License (http://creativecommons.org/licenses/by/3.0), which permits unrestricted use, distribution, and reproduction in any medium, provided the original work is properly cited.

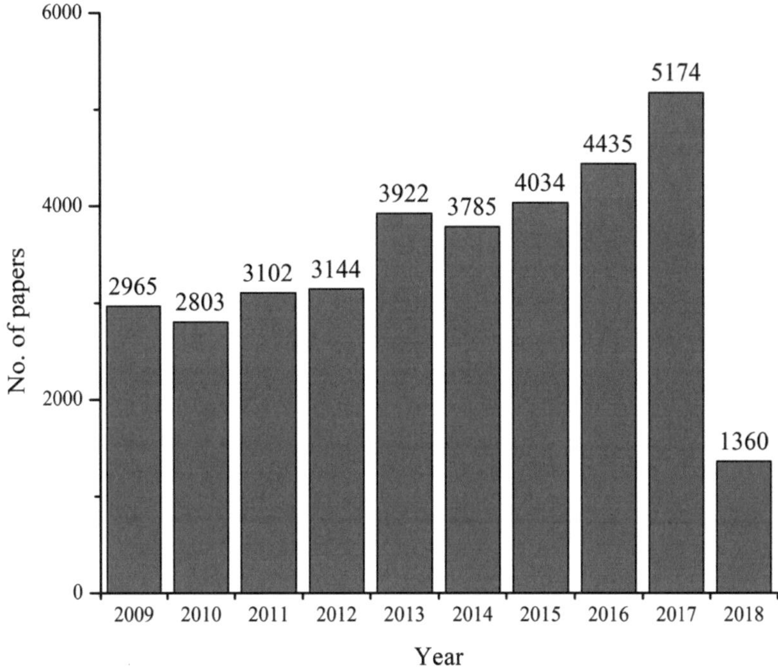

Figure 1. The number of publications with keyword arsenic, according to Science Direct.

environment and can originate from natural and anthropogenic sources [3]. Natural sources of arsenic are: rocks with incorporated arsenic compounds, activity of volcanoes and some biological processes. Anthropogenic sources are numerous, from mining to different types of production (pesticides, wood preservatives, and pigments). When the arsenic compounds reach groundwater, it is hard to distinguish the origin, both natural and anthropogenic arsenic species are released [3].

According to Science Direct, during the last decade, a significant number of scientific papers reporting the results from arsenic investigations are presented in **Figure 1**. The focus of these researches was the development and improvement of methods for arsenic detection, extraction, separation and removal.

The investigation of arsenic species and their behavior in various samples, especially in natural waters and environment is important for chemistry and environmental protection. The most common arsenic species are presented in **Table 1**

Depending on the oxido-reduction conditions, microbiological environment, arsenic species can be present in water in solution or in a precipitated form, and they can also adsorb or desorb from the existing precipitates [1, 2]. When arsenic species are soluble in water, they can be present in both inorganic and organic forms. For iAs species both As(III), arsenite, and As(V), arsenate, can be present. For oAs species, MMA and DMA are soluble forms of organic arsenic species. From the value of the chemical equilibrium constants for each molecular or ionic form of arsenic in water, the present species can be recognized [3]. When choosing and

Arsenic species	Oxidation state	Chemical formula	Group	Presence in the environment
As(V)	+5	AsO_4^{-3}	iAs	Water
As(III)	+3	AsO_3^{-3}		
MMA	+5	$CH_3AsO(OH)_2$	oAs	
DMA	+5	$(CH_3)_2AsO(OH)$		
TMAO	+5	$(CH_3)_3AsO$		Seafood (fish, mussels)
TETRA	+3	$(CH_3)_4As^+$		
AsB	+3	$(CH_3)_3As^+CH_2COO^-$		
AsC	+3	$(CH_3)_3As^+CH_2CH_2OH$		

Table 1. Common inorganic and organic arsenic species [5].

analyzing the most dominant form of arsenic in water, the most present is inorganic arsenic as As(V). If As(III) is present, there are two important things that need to be taken into account. As(III) is more poisonous (even at low concentrations) than As(V). Beside the severe toxic effect, As(III) is easily oxidized. In oxidized conditions, stable forms of arsenic are As(V), and MMA and DMA, from oAs species. Many water sources in the world containing high concentration of arsenic cause health problems or diseases such as cancer. The WHO provisional guideline value for arsenic in drinking water is 10 μg L^{-1} [4]. Water quality analysis usually do not include test on arsenic. Arsenic compounds are colorless and odorless.

Once the presence of arsenic is determined in water, the separation and removal is obligatory. Removal technologies that are efficient, but still need improvement include absorption, precipitation, different electrochemical processes, membrane and hybrid membrane processes [6–9].

2. Arsenic in water

Arsenic enters the water through the dissolution of minerals, ores soil, sediments, water, living organisms and rocks containing high concentrations of arsenic. Drinking water from surface water bodies usually does not contain high concentrations of arsenic. Higher concentrations have only been found in the groundwater. Human activities influence and change the content of arsenic in nature. When using arsenic compounds for different purposes, there is a direct influence. There is also indirect influence that affects the mobility of arsenic from different natural sources. Organic arsenic compounds such as AsB, AsC, TETRA, TMAO, arsenosugars and arsenic-containing lipids are mainly found in marine organisms although some of these compounds have also been found in terrestrial species.

Despite the fact that iAs species are predominant in natural waters, the presence of oAs has also been reported. Even though the main analytical interest is to determine total arsenic in water, it is also important to develop the procedures for As species determination, separation, and removal. The distribution of i As and oAs species is a function of pH value of water [2].

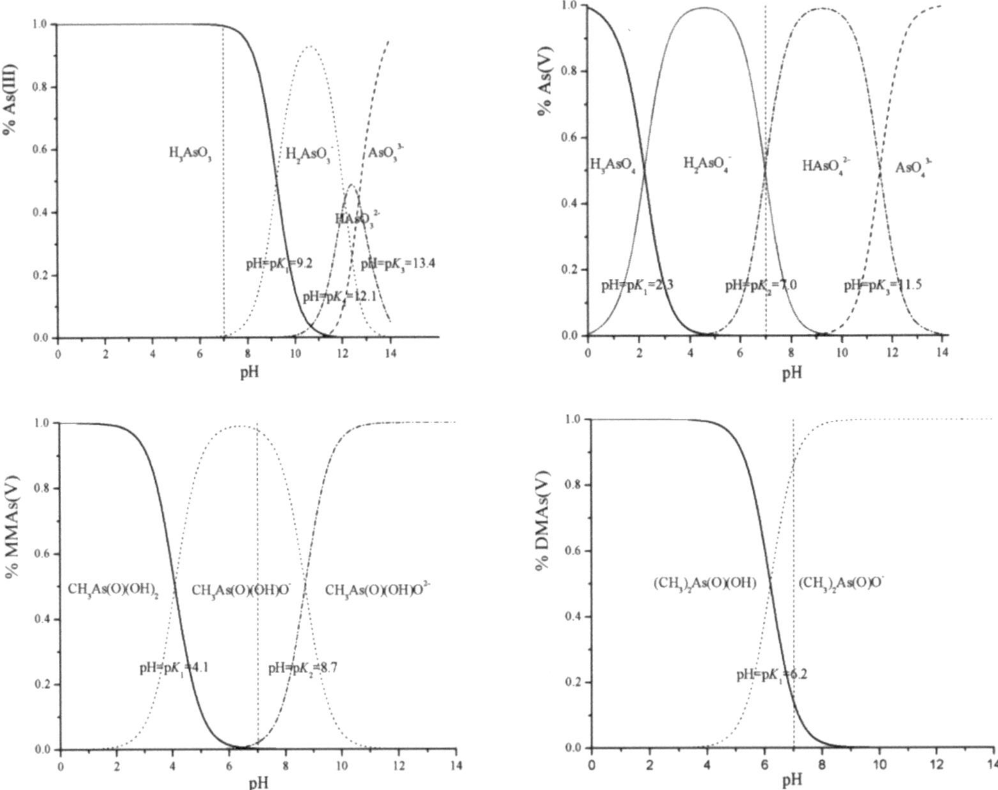

Figure 2. The distribution of iAs and oAs species as a function of pH values of water [2]. Copyright approved by publisher.

The distribution of arsenic species vs. pH values of water is presented in **Figure 2** [2].

As(III) species: H_3AsO_3, $H_2AsO_3^-$, $HAsO_3^{2-}$ and AsO_3^{3-}, are stable under slightly reducing aqueous conditions. As(V) species: H_3AsO_4, $H_2AsO_4^-$, $HAsO_4^{2-}$ and AsO_4^{3-}, are stable in oxygenated waters [6]. Two valences of the same element, molecular (ortho, H_3AsO_3, H_3AsO_4 and meta forms, $HAsO_2$, $HAsO_3$) and ionic forms with different charges make the research of arsenic removal from water more challenging and indivisible of arsenic chemistry knowledge. Any arsenic removal technology strongly depends on the water conditions and the stability of arsenic forms in the water.

Bearing in mind the fact that arsenic occurs in water in molecular and ionic form depending on water pH, the main goal of many investigations is to select the most efficient exchanger, not only in terms of efficiency, but also in terms of applicability in the wide range of water pH values in real and environmentally friendly water treatment systems. In neutral conditions, As(V) species are completely in ionic form ($H_2AsO_4^-$ and $HAsO_4^{2-}$), while As(III) is in molecular (H_3AsO_3 or $HAsO_2$), as shown in **Figure 2** [2].

3. Determination of arsenic and arsenic species in water

There are a variety of chemical methods from classical to contemporary analytical techniques that are used for determination of arsenic and arsenic species in water.

There has been several review articles on the speciation of arsenic in a variety of samples [10–14]. These reviews focus on (1) determination of total content of arsenic and (2) speciation analysis.

A review of contemporary methods for arsenic and arsenic species in water is presented in **Table 2**. The parameters, as detection limit, advantages and disadvantages are pointed out in order to have an insight into ability and application of available techniques.

The total concentration of arsenic in drinking water (mostly traces of arsenic, level of μg L^{-1} or less) can be detected only by sophisticated analytical techniques as ICP-MS, GF-AAS and HG-AAS [3, 14]. For As speciation analysis, well-established methods that involve the coupling of separation techniques, such as HPLC with a sensitive detection system, that is, ICP-MS, are recommended, and they are mostly used [13].

Methodology	Detection	Detection limit (μg L^{-1})	Advantages	Disadvantage	Ref.
ICP-AES	Total arsenic	~30	Minimal sample volume; no sample pretreatment and short measurement time	Expensive; needs lot of knowledge for operating and interpretation of data	[14]
ICP-MS	Total arsenic	~0.1	Approved by US EPA	Spectral and matrix interferences	[11, 13, 19]
GF-AAS	Total arsenic	~0.025	Approved by US EPA	—	[3, 14]
HG-AAS	Total arsenic and arsenic speciation	0.6–6.0	Approved by US EPA	—	[14]
HPLC-HG-AAS	Total arsenic and arsenic speciation	1–47	No need for sample pretreatment	—	[3, 14]
HPLC-HF-AAS	Arsenic speciation	0.05–0.8	Rapid, inexpensive. No need for sample pretreatment	—	[3, 14]
IC-ICP-MS	Arsenic speciation	0.01	No need for sample pretreatment	—	[19]
HPLC-ICP-MS	Total arsenic	0.01	No need for sample pretreatment	—	[13]

Table 2. A review of contemporary methods for arsenic and arsenic species determination in water.

Historically, *colorimetric/spectrophotometric methods* have been used to determine total arsenic concentration. Several commercial field kits have been based on Marsh and Gutzeit reaction. All As species in a sample reduce to As (arsenic mirror) or arsine, AsH_3, (it passes on to an $HgBr_2$-impregnated filter, turning it to yellow to brown color, depending on the amount of arsenic present). These tests are obvious, visible proofs for arsenic detection, and they are popular and useful in the field of forensic toxicology. The colorimetric methods are easy to use and inexpensive in terms of equipment and operator cost. They are useful for the semi-quantitative determination of high concentrations of arsenic in water. Spectrophotometric methods are based on conversion of arsenic to the colored compound such as molybdenum blue, or silver diethyldithiocarbamate [15, 16].

Electrochemical methods, particularly voltammetric methods, are affordable, sensitive and ease of fabrication, and they are noteworthy for arsenic determination. Much work has been done in this area [12]. The ASV methods using platinum and gold electrodes, and CSV method using a glassy-carbon electrode have very low detection limit for arsenic determination. Determination of total As is performed by reducing As(V) to As(III) using various chemicals, and the limits of detection achieved were in vicinity of 0.02 μg L^{-1}. Also, arsenic in drinking water can be measured with Cu(II) by differential pulse cathodic stripping voltammetry (DPCSV) using hanging mercury drop electrode (HMDE) as working electrode and Ag/AgCl as reference electrode [12, 17, 18].

At present, for total As concentration determination, laboratories often prefer more sensitive methods such as AAS, AES, MS or AFS. Usually, the total concentration of arsenic needs to be determined, then the speciation analysis follows.

To perform speciation analysis properly, the best option is coupling of two analytical techniques. One technique is used for the separation of all chemical forms of arsenic that are present in water, and the other is used for the detection of these species. Besides coupling analytical techniques, there are necessary steps for complete analysis of arsenic. The first one is the extraction of arsenic, which has to be both mild and effective, at the same time. The second step is separation of various forms of arsenic species. The final step is the measuring step which gives the answer to the quantification of each present arsenic compound.

3.1. Sophisticated coupling technique

Analytical methods for determining different arsenic species have become increasingly important due to different toxicity and chemical behavior of various arsenic forms. Methods that involve the coupling of separation techniques, such as IC and HPLC with a sensitive detection system, such as ICP-MS, HG-AFS, HG-AAS and GF-AAS [3, 11, 13, 14, 19]. HPLC has been a preferred technique used for separation of arsenic compounds. Coupled with ICP-MS for determination, as HPLC-ICP-MS system it is a method of choice for separation and measurements all arsenic species in water. In addition, applying IC coupled with ICP-MS, it is possible to separate and estimate arsenic species in water: iAs(III), iAs(V), DMA, MMA, AsBet. A representative result is presented in **Figure 3** [19].

The evaluation of analytical method is based on defining: selectivity, repeatability, accuracy, specific features of the method and defining the limits of detection and quantification (LoD

http://dx.doi.org/10.5772/intechopen.75531

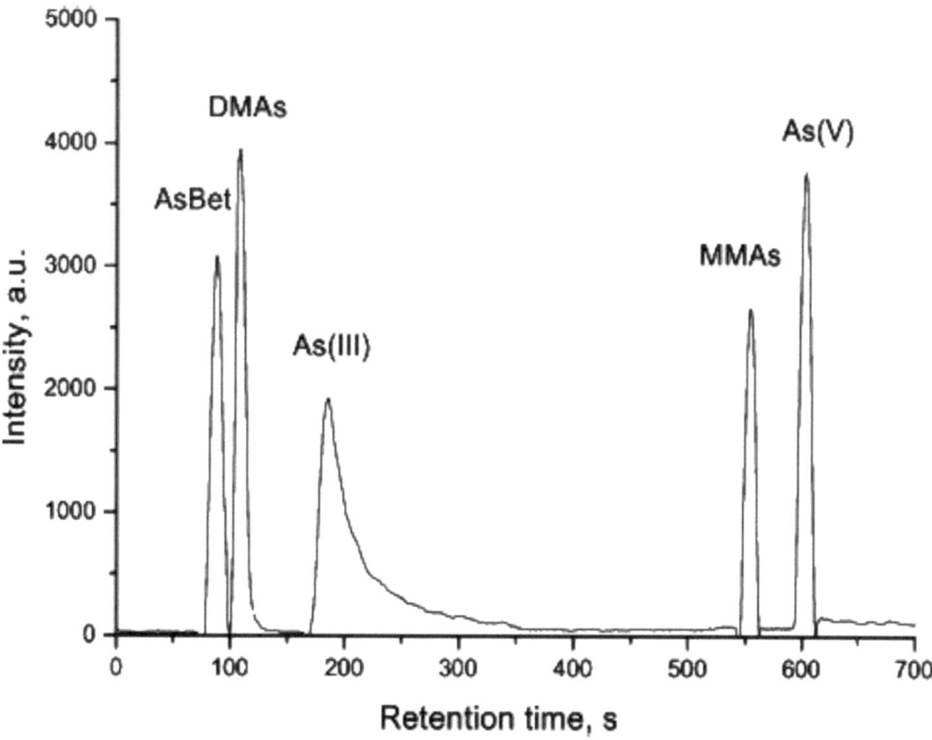

Figure 3. Determination of five arsenic species by IC-ICP-MS. Mobile phase: NaOH [19].

and LoQ). These limits, these numbers give the information on the smallest concentration that can be detected and quantified with certain accuracy that has been defined [10]. The LoD was discussed and determined for the induced coupled plasma-mass spectrometry (ICP-MS) measurements of arsenic [11]. Thorough analysis has shown that the best option for LoD would be experiments, which would include the repetition many times. If experiments would be repeated 100 times, it is expected that only five measurements would be inadequate. Although this is ideal, the time consumption for the repetitive measurements is not acceptable. The most important conclusions were that LoD is not permanent and constant value, and it has to be verified and adopted for each new case. LoD is a basic parameter for estimation of the LoQ. It was concluded in [11] that the traditional (IUPAC) method is the one that could be applied.

4. Removal of arsenic and arsenic species from water

Different methods can be applied for arsenic removal from water. Arsenic (V) is more effectively removed than As(III) by both conventional and nonconventional methods. Pretreatmen (preoxidation) of As(III) to As(V) is an essential step for better removal [2]. Methods that have been successfully applied in water treatment plants are: precipitation and coprecipitation, electrochemical (such as electrocoagulation), ion exchange and MST (reverse osmosis, ultrafiltration and other membrane techniques) [6–9, 20, 21].

4.1. Sorption processes for arsenic separation and removal

A wide range of sorbent materials for aqueous arsenic removal has been tested and used: biological materials, mineral oxides, activated carbons and polymer resins. Even some agricultural and industrial by-products such as red mud, fly ash, waste iron slag from steel production plant and waste filter sand from water treatment plant, have proved to be good and inexpensive arsenic sorbents [6, 7]. The potential use and application of industrial wastes in water treatment is in favor of the eco-friendly concept that preserves natural resources and supports the reuse-recycle concept. The technology of arsenic adsorption is based on materials which have a high affinity for dissolved arsenic. Adsorption of arsenic by iron modified sorbents has been established by several authors [6, 7]. There are numerous scientific and professional investigations with intention to develop a small and efficient system for arsenic removal based on natural and artificial sorption materials [20, 21]. Large amount of chemicals used for *precipitation and coprecipitation* processes (alum sulfate or ferric chloride) produce sludge, which needs treatment before disposal. If not treated properly, leachate with high concentration of arsenic is emitted to soil, threatening to contaminate the aquifers.

A step forward has been made by investigations that were devoted to the evaluation of selective multifunctional sorbents including ion-exchange resins for SPE and chromatographic columns connected with a sensitive measurements system [2]. The need to determine As species in water resulted in developing new materials for arsenic separation and removal. A simple procedure for selective separation (in pretreatment) of arsenic species in water using chemically modified and unmodified ion-exchange resins is presented in **Figure 4** [2].

For separation of As species in water, two types of resins, strong base anion exchange resin (SBAE), hybrid resins (HY) and hybrid resin chemically modified (HY-Fe and HY-AgCl), were tested and used. The HY-Fe resin retained all arsenic species except DMAs(V). This is recognized as an advantage because this makes direct measurement of this species in the effluent possible. The HY-AgCl resin retained all iAs, which was convenient for direct determination of oAs species in the effluent. The selective bonding of arsenic species on three types of resins, as shown in **Figure 4**, has been established as the procedure which enables the separation and calculation of all arsenic species in water [2].

EC comprises complex chemical and physical processes involving many surface and interfacial phenomena. Very effective and perspective EC process consists of three processes: electrochemical reactions (simultaneous anodic oxidation and cathodic reduction), flotation and coagulation [9, 20]. The EC process relies on the generation of metal ions from electrodes. The electrodes can be made of iron, aluminum or zinc, depending on the most favorable reactions for arsenic removal. The reaction in reaction chamber starts after the application of direct current. The electrode (metallic anode) dissociates into valent metallic ions. The metallic ions migrate to oppositely charged ions and the precipitation of different insoluble salts occur (different sulfides, oxides, hydroxides, chromates or phosphates, depending on the presence of ions in water). EC has several advantages when compared to other methods. The construction of reaction chamber is compact, control of the process is simple, no additional chemicals are required, and the result is reduced amount of sludge. If the electrode is made of iron, ferric hydroxide is one of the main solid products, as shown in Eq. (1) [9]:

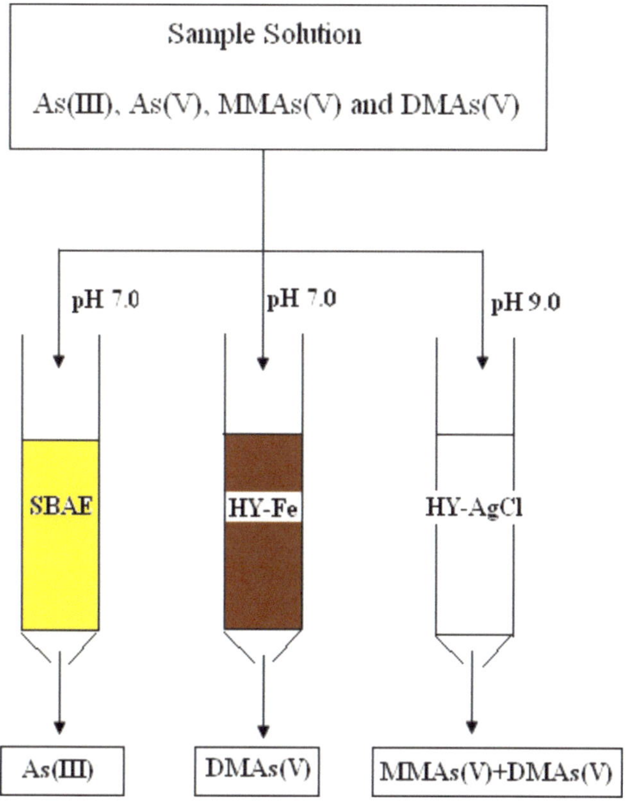

Figure 4. Procedure for selective separation arsenic species in water using ion-exchange resins [2]. Copyright approved by publisher.

$$Fe^{3+}(aq) + 3OH^{-}(aq) \rightleftharpoons Fe\,(OH)_3(s). \tag{1}$$

Arsenate co-precipitates or adsorbs to $Fe(OH)_3(s)$, as shown in Eq. (2).

$$Fe\,(OH)_3(s) + AsO_4^{3-}(aq) \rightleftharpoons \left[Fe\,(OH)_3 \cdot AsO_4^{3-}\right](s). \tag{2}$$

The potential of EC as an alternative water treatment technique to remove arsenic from water needs to be realized [8, 9, 20].

Ion-exchange, IE, processes with regeneration capability is a proven, efficient and low-cost treatment method for the exchange of arsenic in the As(V) form [1, 2]. The ion-exchange reaction between As(V) and a bed of chloride-form SBAE resin (designated as R-Cl resin) occurs as presented by Eq. (3):

$$2\,R\text{–}Cl + HAsO_4^{2-} \rightleftharpoons R_2 - HAsO_4 + 2Cl^{-}. \tag{3}$$

When the regeneration of resins is needed, both HCl and NaCl can be applied. Still, with HCl solution, more efficient regeneration occurs because the ionic forms of arsenic (anions)

Technology for arsenic removal	Advantage	Disadvantage	Some specific feature	Future perspective
Adsorption	Cheap materials, effective and efficient removal	Further treatment for regeneration and consumption of chemicals	Additional filter for removal of fine particles is required	Still attractive as an efficient and cheap technology for As removal. Finding new, environmentally friendly sorbent is still a challenging task
Chemical coagulation	Effective for industrial wastewater treatment plants and efficient for As(V) removal	Chemical required. pH adjustment needed. Large volumes of sludge that needs further treatment	Arsenic leaching out from sludge	Not attractive as a solution, only if it coupled with electrochemical techniques
Electrocoagulation	Efficient for arsenic removal. Low maintenance costs. No chemicals or pH adjustment. Low operating costs	Applicable only on batch scale. Passive oxide films for on the electrode. High energy consumption	No generation of secondary pollutants	Attractive for future investigations. Need to overcome the lack of application on a large scale
Ion exchange	Efficient for As(V) removal. Exchange resins are available; the selective resins for removing arsenic are one of the most important requirements to provide high removal. Together with hybrid solution is an excellent technology	Interference with other ions. Easily blocked. Huge amount of chemicals	Using this kind of technique depends on the pH values of water	Attractive only if selective and sensitive chemical agents are included in ion-exchange process
Membrane technologies	Efficient in arsenic removal. No chemical reagents. No sludge. Small dimensions for membrane treatment plant. Easy automation and control	Removal of arsenic depends on the pressure, pH value, solute concentration, temperature of feed solution	Arsenic is concentrated in the retentate	Attractive in future perspective. With decrease of investment the MST will prevail in arsenic removal technologies. Different membrane materials and processes need to be evaluated to select the optimum for each situation

Table 3. The comparison and future perspective of different technologies for arsenic removal.

transform to molecular form (H_3AsO_4). Molecular forms do not affect the equilibrium of ion-exchange processes as illustrated by Eq. (4):

$$R_2 - HAsO_4 + 2Cl^- + 2H^+ \rightleftharpoons 2R\text{–}Cl + H_3\,AsO_4. \quad (4)$$

Different sorption processes, from adsorption, to chemisorption and ion-exchange, have shown a potential being efficient and cheap (depending on the selected sorbent). With improved, more selective and chemically modified sorbents, the extraction technique can be replaced [17–19]. What has been specifically used as an advantage for arsenic species separation is different behavior of arsenic species at various pH values [3, 22].

The hybrid resin (HY) that has successfully been applied uses the activity of the hydrated iron oxides (HFO) and anion exchange for selective separation of arsenic [2]. With integrated use of anion exchange and sorption, the separation of As(III) and As(V) species and removal of all species of arsenic can be accomplished. With application of HY resin, two separate things can be accomplished: the collection and preconcentration of low concentrated iAs or the removal of iAs species, if it is interfering the determination.

Membrane separation technologies, such as RO, NF, UF, MF, can be employed in the removal of arsenic from water. Depending on the removal efficiency, RO and NF are more efficient than UF and MF. Operating conditions, membrane material, water quality, temperature, pressure, pH value and chemical compatibility have to be considered during operation of a membrane plant. When MF and UF are applied, less amounts of chemicals are used, and therefore, less sludge is produced. When RO and NF are used, no chemicals are needed and the amount of sludge is neglectable [8].

The comparison and future perspective of different technologies for arsenic removal are presented in **Table 3**.

5. Conclusion

Arsenic contamination of water has been reported as a critical issue in many articles, which reflects the latest state-of-the-art understanding of the behavior and toxicity of various arsenic species. Many water sources in the world contain low concentration of arsenic (mostly traces of arsenic, level of $\mu g\ L^{-1}$ or less). If the concentration of arsenic in drinking water is higher than 10 $\mu g\ L^{-1}$, which is the WHO provisional guideline value for arsenic, it causes various health problems. All arsenic compounds dissolved in water are toxic. In natural waters, arsenic appears most often in inorganic forms and to a lesser extent in organic form. Inorganic species, arsenic acids (H_3AsO_3 and H_3AsO_4) and their ions are more toxic than organic forms. In addition, As(III) species are more toxic than As(V) ones. The valence (+III and +V), the type of arsenic species, ionic or molecular forms are dependent on the oxidation–reduction condition and pH of the water. Arsenic in water occurs in both inorganic and organic forms, but inorganic species are predominant in natural waters. In neutral conditions, As(V) species are completely in ionic form ($H_2AsO_4^-$ and $HAsO_4^{2-}$), while As(III) is in molecular form (H_3AsO_3 or $HAsO_2$).

Arsenic compounds are colorless and odorless, and testing water for arsenic is an important strategy for the health and well-being of people. Working with a water professional to monitor and maintain the quality of the well and water supply is an important responsibility.

In this work, methods for arsenic and arsenic speciation separation, determination and removal were reviewed. There are numerous methods for separation and determination of arsenic species in water. It is very important to recognize easy, simple and inexpensive methods to estimate the very low concentrations of arsenic.

The total concentration of arsenic in drinking water can be detected by simple Gutzeit method, and some similar colorimetric methods of comparing stains produced on treated paper strips. Although its minimum detectable concentration is 1.0·μ L^{-1}, these tests should be used when only a qualitative or semiqualitative detection is needed.

For precise, and reliable determination of arsenic in water, only sophisticated analytical techniques as ICP-MS, GF-AAS and HG-AAS can be applied. These methods are approved by US EPA. The features of these methods are high sensitivity, high accuracy, minimal sample volume; no sample pretreatment and short measurement time with minimum detectable concentration of 0.1 μ L^{-1}. They are expensive, need lot of knowledge for operating and interpretation of data.

For As speciation analysis, well-established methods that involve the coupling of separation techniques, such as HPLC with a sensitive detection system, that is, ICP-MS, are recommended, and they are mostly used. Through the limits, it is possible to define the smallest concentration of analyte that can be reliably detected and quantified. Limit of detection for the HPLC-ICP-MS system is 0.001 μ L^{-1}. This system is also expensive and needs lot of knowledge for operating and interpretation of data.

In all works, a special attention is paid to the preservation of arsenic species in environmental water samples for reliable speciation analysis. An appropriate procedure for the extraction of arsenic species from water should be accomplished without changing any original state of arsenic. This is still a challenging topic for research. The proposed system showed themselves to be accurate, precise and time efficient, as just a very simple sample treatment is required. Successful application of all methods required considerable practice.

Sorption processes (ion exchange, adsorption, chemisorption) with regeneration capability are proven as efficient and low-cost treatment methods for the removal of arsenic species from water. Separation of arsenic species using these new selective and chemically active sorbents recognize as a cost- and time-saving alternative to the traditional extraction techniques. The major drawback of all these techniques is that they are unable to remove As(III) efficiently.

Membrane separation technologies, such as RO, NF, UF, MF, are recommended for the removal of arsenic from water in water treatment plants.

Although there are numerous research papers focused on extraction techniques, yet it is not possible to set universal extraction procedures. These procedures depend on the presence of different species as well as on the type of matrices. For arsenic speciation, the choice of the most appropriate method is of great importance for obtaining reliable and accurate results.

Acknowledgements

The authors are grateful to the Ministry of Education and Science of the Republic of Serbia which supported our scientific work (projects no. TR37009, TR37010 and III43009).

Abbreviations

Arsenic compounds

As	arsenic
iAs	inorganic arsenic
oAs	organic arsenic
As(III)	arsenite ion
As(V)	arsenate ions
MMA	monomethylarsenic acid
DMA	dimethylarsenic acid
TMAO	trimethylarsine oxide
TETRA	tetramethylarsonium ion
AsB	arsenobetaine
AsC	arsenocholine

Methods and techniques for arsenic determination

IC	ion chromatography
HPLC	high-performance liquid chromatography
MS	mass spectrometry
AES	atomic emission spectrometry
ICP-MS	inductively coupled plasma-mass spectrometry
ASV	anodic stripping voltammetry
CSV	cathodic stripping voltammetry

DPCSV	differential pulse cathodic stripping voltammetry
GF-AAS	graphite furnace absorption spectrometry
HG-AAS	hydride generation atomic absorption spectrometry
HPLC-HG-AAS	high-performance liquid chromatography-hydride generation-atomic absorption spectrometry
HPLC-HG-AFS	high-performance liquid chromatography or solid-phase cartridge separation combined with hydride generation-atomic fluorescence spectrometry
HPLC-ICP-MS	high-performance liquid chromatography-inductively coupled plasma-mass spectrometry

Methods and techniques for arsenic removal

SPE	solid phase extraction
IE	ion exchange
SBAE	strong base anion exchange resin
HY	hybrid resin
EC	electrocoagulation
RO	reverse osmosis
NF	nanofiltration
UF	ultrafiltration
MF	microfiltration
MST	membrane separation technologies

Author details

Ljubinka Rajakovic[1]* and Vladana Rajakovic-Ognjanovic[2]

*Address all correspondence to: ljubinka@tmf.bg.ac.rs

1 Faculty of Technology and Metallurgy, University of Belgrade, Belgrade, Serbia

2 Faculty of Civil Engineering, University of Belgrade, Belgrade, Serbia

References

[1] Issa N, Rajaković-Ognjanović V, Jovanović B, Rajaković L. Determination of inorganic arsenic species in natural waters-Benefits of separation and preconcentration on ion exchange and hybrid resins. Analytica Chimica Acta. 2010;**673**(2):185-193. DOI: 10.1016/j.aca.2010.05.027

[2] Issa N, Rajaković-Ognjanović V, Marinković A, Rajaković L. Separation and determination of arsenic species in water by selective exchange and hybrid resins. Analytica Chimica Acta. 2011;**706**(1):191-198. DOI: 10.1016/j.aca.2011.08.015

[3] Rajaković L, Todorović Ž, Rajaković-Ognjanović V, Onjia A. Review: Analytical methods for arsenic speciation analysis. Journal of the Serbian Chemical Society. 2013;**78**:1-32. UDC JSCS-5666

[4] Guidelines for Drinking-Water Quality. Recommendations World Health Organization; WHO Press; Incorporating the First and Second Addenda. Vol. 1, 3rd ed. Geneva, Switzerland: World Health Organization; 2008

[5] Francesconi K, Kuehnelt D. Arsenic compounds in the environment. In: Frankenberger WT Jr, editor. Environmental Chemistry of Arsenic. New York: Marcel Dekker; 2002. pp. 51-94

[6] Lekić B, Marković D, Rajaković-Ognjanović V, Đukić A, Rajaković L. Arsenic removal from water using industrial by-products. Journal of Chemistry. 2013;**2013**:9 pages ID 121024. DOI: 10.1155/2013/121024

[7] Vukašinović-Pešić V, Rajaković-Ognjanović V, Blagojević N, Grudić V, Jovanović B, Rajaković L. Enhanced arsenic removal from water by activated red mud based on hydrated Iron(III) and titan(IV) oxides. Chemical Engineering Communications. 2012;**199**(7):849-864. DOI: 10.1080/00986445.2011.631235

[8] Uddin T, Mozumder S, Figoli A, Islam A, Drioli E. Arsenic removal by conventional and membrane technology: An overview. Indian Journal of Chemical Technology. 2007;**14**:441-450

[9] Song P, Yang Z, Zeng G, Yang X, Xu H, Wang L, Xu R, Xiong W, Ahmed K. Electrocoagulation treatment of arsenic in wastewaters: A comprehensive review. Chemical Engineering Journal. 2017;**317**:707-725. DOI: 10.1016/j.cej.2017.02.086

[10] Ariño C, Serrano N, Díaz-Cruz JM, Esteban M. Voltammetric determination of metal ions beyond mercury electrodes. A review. Analytica Chimica Acta. 2017;**990**:11-53. DOI: 10.1016/j.aca.2017.07.069

[11] Rajaković L, Marković D, Rajaković-Ognjanović V, Antanasijević D. Review: The approaches for estimation of limit of detection for ICP-MS trace analysis of arsenic. Talanta. 2012;**102**:79-87. DOI: 10.1016/j.talanta.2012.08.016

[12] Liu Z-G, Huang X-J. Voltammetric determination of inorganic arsenic (A review). Trends in Analytical Chemistry. 2014;**60**:25-35. DOI: 10.1016/j.trac.2014.04.0140165-9936

[13] Seulki J, Hosub L, Youn-Tae K, Hye-On Y. Development of a simultaneous analytical method to determine arsenic speciation using HPLC-ICP-MS: Arsenate, arsenite, monomethylarsonic acid, dimethylarsinic acid, dimethyldithioarsinic acid, and dimethylmonothioarsinic acid. Microchemical Journal. 2017;**134**:295-300. DOI: 10.1016/j.microc.2017.06.011

[14] Hung D, Nekrassova O, Compton R. Analytical methods for inorganic arsenic in water: A review. Talanta. 2004;**64**:269-277. DOI: 10.1016/j.talanta.2004.01.027

[15] Rupasinghea T, Cardwella TJ, Cattralla RW, Luquede MD, Spas C, Koleva D. Pervaporation-flow injection determination of arsenic based on hydride generation and the molybdenum blue reaction. Analytica Chimica Acta. 2001;**445**:229-238. DOI: 10.1016/S0003-2670(01)01256-9

[16] Gupta PK, Prabhat KG. Microdetermination of arsenic in water, spectrophotometrically, by arsine-silver diethyldithiocarbamate-morpholine-chloroform system. Microchemical Journal. 1986;**33**:243-251. DOI: 10.1016/0026-265X(86)90064-0

[17] Ma J, Sengupta MK, Yuan D, Dasguptab PK. Speciation and detection of arsenic in aqueous samples: A review of recent progress in non-atomic spectrometric methods. Analytica Chimica Acta. 2014;**831**:1-23. DOI: 10.1016/j.aca.2014.04.029

[18] Guo Z, Yang M, Huang X. Recent developments in electrochemical determination of arsenic. Current Opinion on Electrochemistry. 2017;**3**:130-136. DOI: 10.1016/j.coelec.2017.08.002

[19] Vassileva E, Becker A, Broekaert J. Determination of arsenic and selenium species in groundwater and soil extracts by ion chromatography coupled to inductively coupled plasma mass spectrometry. Analytica Chimica Acta. 2001;**441**:135-146. DOI: 10.1016/S0003-2670(01)01089-3

[20] Nidheesha P, Anantha Singh T. Arsenic removal by electrocoagulation process: Recent trends and removal mechanism. Chemosphere. 2017;**181**:418-432. DOI: 10.1016/j.chemosphere.2017.04.082

[21] Rajaković L, Mitrović M. Arsenic removal from water by chemisorption filters. Environmental Pollution. 1992;**75**(3):279-287. DOI: 10.1016/0269-7491(92)90128-W

[22] Syahidah N, Zulkifli H, Abdul R, Woei-Jye L. Detection of contaminants in water supply: A review on state-of-the-art monitoring technologies and their applications. Sensors and Actuators B: Chemical. 2018;**255**:2657-2689. DOI: 10.1016/j.snb.2017.09.078

Chapter 3

Bio-adsorbents for the Removal of Heavy Metals from Water

Khathutshelo Catherine Mqehe-Nedzivhe,
Khathutshelo Makhado,
Oluwasayo Folasayo Olorundare,
Omotayo Ademola Arotiba, Elizabeth Makhatha,
Philiswa Nosizo Nomngongo and
Nonhlangabezo Mabuba

Additional information is available at the end of the chapter

http://dx.doi.org/10.5772/intechopen.73570

Abstract

The work represents the bio-adsorption of arsenic(III) from standard solutions and real water samples using a powdered avocado seed as a bio-adsorbent. The adsorbent was synthesized, demineralized, and characterized by X-ray diffraction (XRD), scanning electron microscope coupled with energy dispersive spectroscopy (SEM-EDS), Fourier transformation infrared spectroscopy (FTIR), and Brunauer-Emmett-Teller (BET) theory. Batch adsorption studies were carried out by using avocado seed, and As^{III} was analyzed by using inductively coupled plasma optical emission spectroscopy (ICPOES) after optimizing the following parameters: pH 6, analyte concentration 2 mg L^{-1}, bio-adsorbent dosage 0.8 g, contact time 120 min between analyte and adsorbent, and temperature from 22 to 40°C. The adsorption capacity of 93.75 mg/g was obtained, and the Langmuir isotherm was adopted by the adsorbent due to the chemisorption that occurs on the surface between the functional groups of the bio-adsorbent and As^{III}.

Keywords: arsenic(III), bio-adsorption, inductively coupled plasma optical emission spectroscopy

© 2018 The Author(s). Licensee IntechOpen. This chapter is distributed under the terms of the Creative Commons Attribution License (http://creativecommons.org/licenses/by/3.0), which permits unrestricted use, distribution, and reproduction in any medium, provided the original work is properly cited.

1. Introduction

Arsenic (As) is the 20th most abundant element in the earth's crust, and its concentration on the soil level is about 5–13 mg/kg [1]. Naturally, arsenic is found in different oxidation states: V (arsenate), III (arsenite), 0 (arsenic), and -III (arsine) [2, 3]. Arsenic (As) is one of the elements that occur naturally and commonly found as an impurity in metal ores, and it is in abundance. It can be found in the soil, water, and living organisms. It is produced commercially for use in wood preservatives, metal alloys, and pesticides.

Both As^{III} and As^{V} exist in the pH range of 6–9, and when comparing As^{III} and As^{V}, it is known that As^{III} is more toxic than As^{V} and also in terms of mobility [4]. Depending on the pH, Arsenite (As^{III}) exists in four forms in aqueous solution, such as H_3AsO_3, $H_2AsO_3^{-}$, $HAsO_3^{2-}$, and AsO_3^{3-}. In a reductive environment, below pH 9.1 As(III) exists in the form of inorganic arsenite (H_3AsO_3) and is thermodynamically stable [5].

Since arsenic is one the most well-known poisonous elements in the periodic table, it is known to be carcinogenic in many parts of the world [6]. Long-term exposure to arsenic, through water and food, can lead to serious health problems like neurological effects, hypertension and cardiovascular diseases, and skin and lung cancer [7]. The kidney is the major source for regulation of water and electrolytes, waste, and chemical compounds, and arsenic(III) can affect the role of the proximal tubules and glomerulus of the kidney [8].

Arsenic(III) is the one that is known to be toxic [4]. The World Health Organization (WHO) recommended a more rigid limit of 10 μg/L as the maximum acceptable arsenic level [9]. Commercial methods for removing arsenic involve technologies such as precipitation, membranes, and adsorption. But adsorption is the most easy, flexible, inexpensive method to be applied by using mineral oxides [10], polymer resins [11], and activated carbons [12]. According to literature, waste materials such as rice husks, tea, and agricultural waste have been applied as inexpensive adsorbents [13].

It is well known that South Africa is a developing country with limited resources for water purification; therefore, the aim of this work is to develop a cheaper, easy-to-use method of treating water especially in the rural areas where the water treatment stations are not yet established.

The most crucial aim of the work is that the bio-adsorbent must be manufactured from a locally available material that is reusable in order to save the rural area's people from daily expenses.

2. Materials and experimental method

2.1. Reagents and standards

Potassium carbonate and arsenic standard employed in the synthetic application procedures of this work were of analytical grade and obtained from Sigma-Aldrich (St. Louis, MO, USA). Millipore water (Merck, Darmstadt, Germany) of 18 MΩ cm^{-1} was used throughout the experiments. Concentrated nitric acid (HNO_3) (70%) was also purchased from (St. Louis, MO,

USA, www.sigmaaldrich.com). All materials especially plastics and glassware prior to use were cleansed by soaking in dilute HNO_3 and rinsed with copious deionized water. The stock standard solutions of As(III) at a concentration of 1000 mgL^{-1} was obtained by dissolving an appropriate amount of arsenic oxide. The working standards were prepared daily by stepwise dilution of stock solution. Avocado fruit waste seeds (AFWS) were locally sourced from fruit and vegetable shop outlet around the city of Johannesburg (Gauteng Province, South Africa).

2.2. Instrumentation

The As quantification was performed using inductively coupled plasma optical emission spectroscopy (ICPOES) (iCAP 6500 Duo, Thermo Fisher Scientific, UK) equipped with a charge injection device (CID) detector. The sample injection application was done through a concentric nebulizer and a cyclonic spray chamber. A Mettler Toledo pH meter model 120 (Greifensee, Switzerland) was employed for all pH measurements. The active carbon material production, i.e., activation, was performed in a tubular furnace (Gallenkamp, Germany). The AFWS and ACM were characterized for porosity, pore structures, surface area, and pore volume using N_2 gas adsorption BET (Brunauer, Emmett, and Teller) method for surface area, while the crystallinity of the material was determined by X-ray diffraction (XRD, Rigaku, UHMa IV, Japan). Surface morphology and surface functional groups were determined by scanning electron microscopy (SEM, TECAN VEGA 3 XMU, Czech Republic) coupled with energy dispersive spectroscopy (EDS) (TECAN VEGA 3 XMU, Czech Republic) and Fourier transmission infrared spectroscopy (FTIR, PerkinElmer FTIR, UK), respectively.

2.3. Adsorbent preparation and application

The precursor for the active carbon material was acquired by collecting avocado seeds. The avocado fruit waste seeds (AFWS) were air-dried and thoroughly washed before rinsing with deionized water. The washed seed residues were then oven-dried overnight at 100°C. The seeds were then pulverized by ball milling it with laboratory hammer mill (Janke and Kunkel Micro-hammer Mill, Staufen IM Breisgau, Germany) to obtain fine powder. The bulky powdered material was later fractionated to diverse particle size using laboratory sieves. The resultant particles of different miniature diameters ranging from 38 to 150 μm were obtained in which 75 μm particle size was subsequently employed in production of active carbon material. The powdered form of avocado seed was then stored inside desiccator until application.

The adsorption studies were carried out to evaluate the efficiency of the avocado peel (bio-adsorbent) for the removal of As^{III} from the aqueous solution using the batch adsorption method. The batch adsorption experiments were carried out in 50 mL plastic bottles by shaking a constant mass of a predetermined size of adsorbent with arsenic standard solutions. The pH of the solutions was adjusted accordingly by adding either ammonium hydroxide or acetic acid solution. Each flask was sealed and kept in a state of agitation (200 rpm) using a mechanical laboratory shaker for the material to reach equilibrium. Upon equilibrium, the samples were filtered and analyzed using inductively coupled plasma optical emission spectroscopy (ICPOES). Parameters such as pH, concentration of solution, mass of adsorbent, contact time, and temperature were optimized.

The percentage removal of As(III) in solution was calculated using Eq. (1):

$$R\% = \frac{C_0 - C_e}{C_o} \times 100 \tag{1}$$

where R is the percentage (%) removal and Co and Ce are the initial and equilibrium concentrations of the analyte, respectively.

The amount of metal adsorbed by adsorbent was calculated from the difference of metal quantity added to the biomass and metal content of the supernatant (Eq. (2) [14, 15]:

$$q_e = \frac{(C_0 - C_e)}{M} \times V \tag{2}$$

where q_e is the metal uptake (mgg^{-1}), C_0 and C_e are the initial and equilibrium metal concentration (mgL^{-1}), V is the volume of the solution (mL), and M is the mass of the adsorbent (g).

3. Results and discussion

3.1. Characterization of the raw avocado seed and activated carbon material

The FTIR spectrum is an important technique which provides the surface functional groups that significantly contribute in the enhanced adsorption efficiency of the adsorbent. FTIR was used to determine the surface functional groups of raw avocado fruit waste seed. In **Figure 1**, the spectrum of the powdered avocado seed is represented, where the band located at 3259 cm^{-1} corresponds to v (O—H) vibrations in the hydroxyl group, while the strong peaks

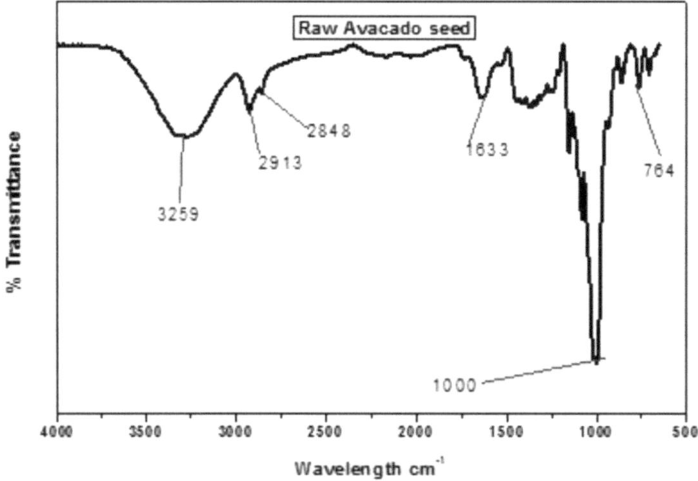

Figure 1. FTIR spectrum of avocado fruit waste seed (AFWS).

at 2913 and 2848 cm^{-1} bands correspond to v(C—H) vibration in the alkane/alkyl aliphatic group which could be methylene [16–20]. The presence of alcohols and carbonyl groups was confirmed by the bond vibrations observed at 1633 cm^{-1}. This confirms that avocado seed is composed of carboxylic group which is responsible for adsorption; similar results were reported in the literature [21, 22].

The surface morphology and the chemical composition of raw avocado seed was studied with scanning electron microscope coupled with energy dispersive spectroscopy (SEM-EDS). The image in **Figure 2(a)** showed that raw avocado seed material had a smooth surface with long ridges and a series of graphitic layers with various pores.

The EDS analyses performed on avocado seed revealed that the surface contained different mineral particles, such as carbon, oxygen, potassium, phosphorus, and chlorine (**Figure 2b**). The highest content of oxygen may assist in adsorption due to electron lone pairs.

3.2. Optimization of adsorption parameters

pH is one of the most important parameters that influence the adsorption of the analyte. In this study, the amount of As^{III} adsorbed on avocado seed was the highest at pH 6 and gradually decreased as the pH increased up to 9 (**Figure 3a**) [7]. However, the highest removal was observed with avocado seed due to the presence of carboxylic group on the surface which increased the affinity toward arsenic to the adsorbent (**Figure 1**). Oxygen of the carbonyl group easily formed the complex with the arsenic [23]. Arsenic(III) adsorption decreased as the pH goes below 6 due to the increasing ionic strength [24].

The effect of the concentration was carried out by increasing the initial concentration from 5 to 30 mg L^{-1}, and the solutions were adjusted to pH 6 at 25°C. It was observed that the percentage removal increased with the increasing concentration of the analyte; this is due to

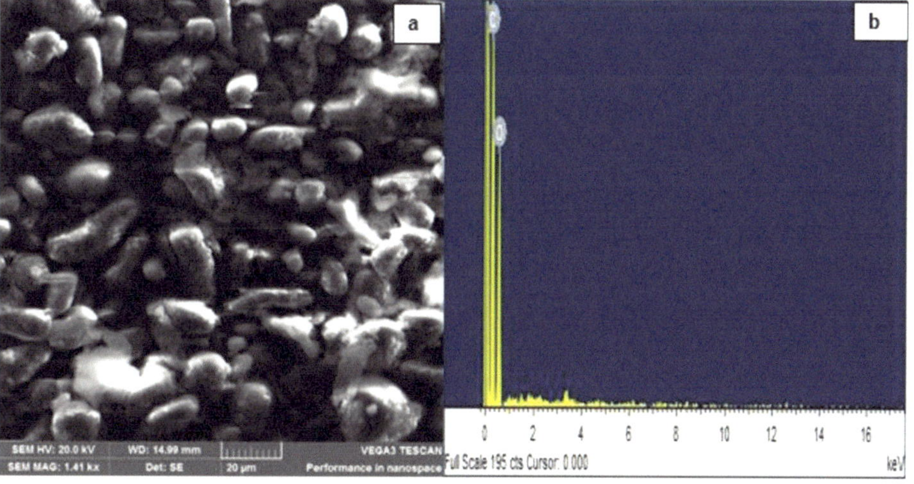

Figure 2. Avocado seed images of (a) SEM and (b) EDS.

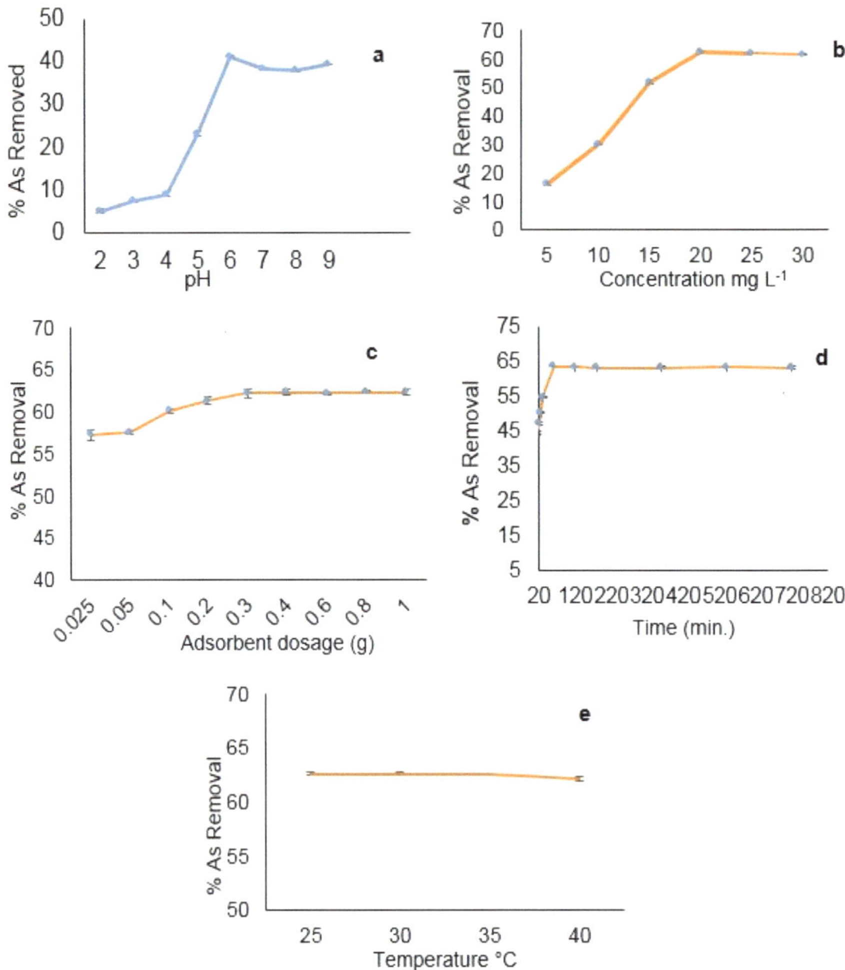

Figure 3. Optimization of (a) pH of the solution, (b) concentration of the analyte, (c) bio-adsorbent dosage, (d) contact time between the bio-adsorbent and the analyte, and (e) temperature of the solution.

the fact that as the concentration increased more ions were available in the solution for adsorption [25]. It was observed from the results in **Figure 3b** that the highest removal (65%) was reached using 20 mg L^{-1} and after there was no significant increase in the percentage (%) removal of As^{III}.

The amount of adsorbent is one on the important factors that affects the adsorption capacity. The adsorbent amount of raw avocado seed on the efficiency of adsorption was investigated, and adsorbent amount was varied from 0.025 to 0.8 g. The results observed indicated that the adsorption increased with increasing adsorbent dosage till 0.8 g (**Figure 3c**). The increase in the percentage removal is due to the availability of active sites for adsorption [26]. It was found that after the dosage of 0.4 g there is no significant change in the percentage removal of arsenic. Then, 0.8 g was used throughout the experiments.

The effect of contact time is an important factor in adsorption because it affects the adsorption kinetics of an adsorbent at the given initial concentration of the adsorbate [26]. The batch adsorption experiments were carried out to investigate the effect of agitation time on the adsorption of As(III). Adsorption rate initially increased rapidly, and the highest removal was reached at 120 min (**Figure 3d**). Further increase in contact time did not show a significant change in the percentage (%) removal of arsenic.

Temperature is one of the parameters that affect the equilibrium and solubility and can also initiate the chemical reaction. This is because temperature can either increase or decrease the activation energy of the analyte. The effect of temperature on the adsorption of arsenic was investigated from room temperature of 40°C. From the results obtained in **Figure 3e**, the temperature did not have any effects since there is no significant increase or decrease in the percentage (%) removal of arsenic.

Under optimized conditions, 2 mg L^{-1} As^{III} standard solution was adsorbed by the avocado peels, and 75% As^{III} was removed (**Figure 4**). The adsorption capacity was 93.75 mg/g when Eq. (2) was applied.

3.3. Adsorption kinetics

Adsorption is described by the functions which connect the amount of adsorbate on the adsorbent.

The distribution of metal ions between the liquid phase and the solid phase is described by several isotherm models such as Langmuir and Freundlich [27].

The Langmuir equation can be written in the form of Eq. (3):

$$\frac{C_e}{q_e} = \frac{1}{q_{\max}} C_e + \frac{1}{K_L q_{\max}} \tag{3}$$

where C_e is the equilibrium concentration (mg/L), q_e is the amount of arsenic adsorbed onto the solid phase (mg/g), b is the equilibrium adsorption constant related to the affinity of binding

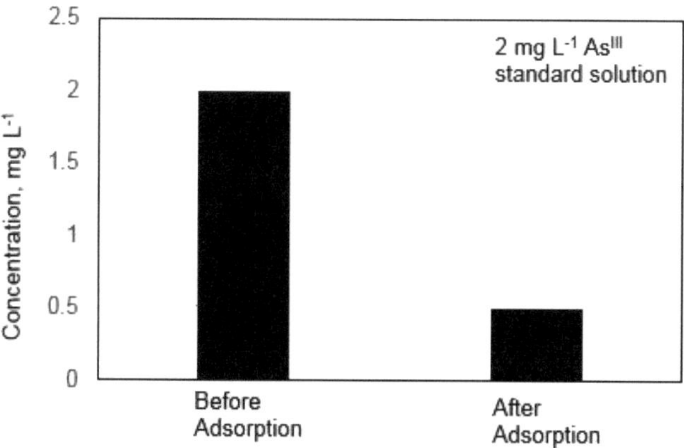

Figure 4. Determination of percentage (%) removal from 2 mg L^{-1} As^{III} standard solution by ICPOES.

sites (L/mg), and q_{max} is the maximum amount of arsenic per unit weight of adsorbent for complete monolayer coverage.

Freundlich equation is represented as shown in Eq. (4):

$$\text{Log}q_e = \log K_f + \frac{1}{n}\log C_e \tag{4}$$

where C_e is the equilibrium concentration (mg/L), q_e is the amount of arsenic adsorbed onto the solid phase (mg/g), K_f is an indicator of the adsorption capacity, and n is the heterogeneity factor.

The results in **Figure 5(a** and **b)** and **Table 1** showed that the correlation coefficient for linear Langmuir model (R^2, 0.97) was higher than the Freundlich model (R^2, 0.72). The data was best fitted in Langmuir model, and this signified that the adsorbent had high affinity for arsenic(III)

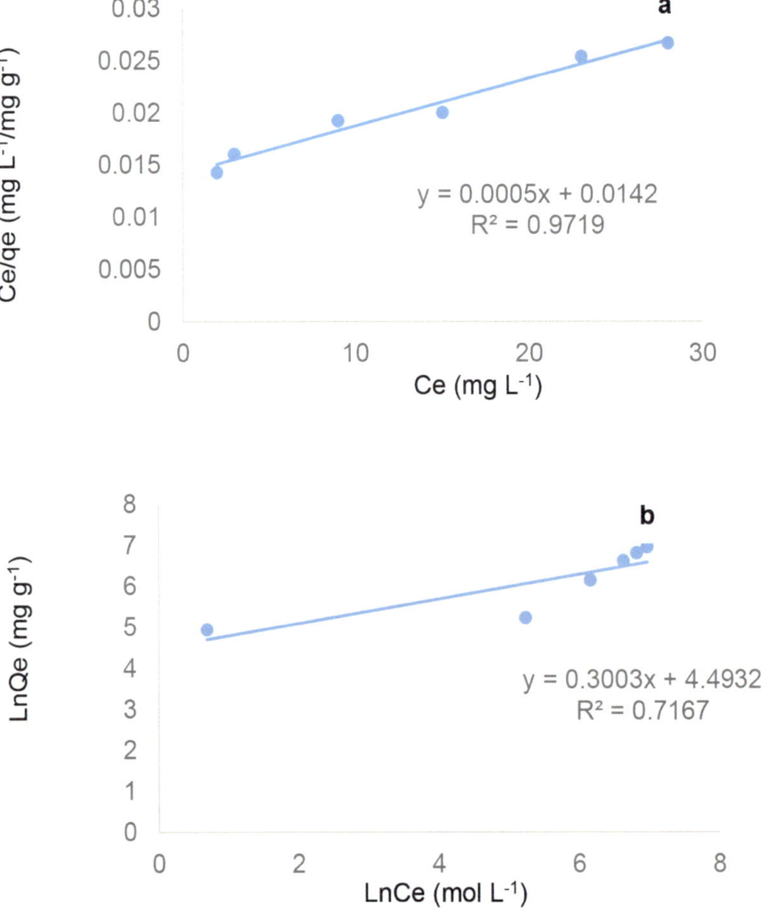

Figure 5. (a) Langmuir isotherm displaying the adsorption of As^{III} onto the surface of the avocado seed by plotting C_e/q_e against equilibrium concentration (C_e). (b) Freundlich isotherm showing the adsorption of As^{III} onto the surface of the avocado seed by plotting ln C_e against equilibrium concentration ln q_e.

due to the carboxylic groups that are the surface of the avocado seed [28], meaning that the chemisorption took place. To prove that the data belonged to Langmuir isotherm, the separation value, R_L value from Eq. (5) was calculated. R_L proves whether the Langmuir adsorption nature is favorable if $R_L > 0$, unfavorable if $R_L > 1$, and irreversible if $R_L = 0$:

$$R_L = \frac{1}{1 + (1 + K_L \times C_0)} \quad (5)$$

where C_0 is the initial concentration and K_L is a Langmuir constant obtained from plotting $1/q_e$ versus $1/C_e$. The results in **Table 1** indicated that the equilibrium sorption was favorable for Langmuir isotherm.

3.4. Analytical figures of merit

Analytical figures of merit for the quantitative analysis of arsenic(III) such as limit of detection (LOD), limit of quantification (LOQ), correlation coefficient (R^2), and the relative standard deviation (RSD) were calculated. In order to determine the LOD, the blank solution was subjected to the optimum experimental conditions, and the signals for ten blank samples were measured (n = 10). The limit of detection (LOD), calculated based on 3S/m (where S is the standard deviation of the blank and m is the slope of the calibration curve) was 0.10 mg L^{-1}. The limit of quantification (LOQ = 10S/m) was 0.20 mg L^{-1} for arsenic(III). The linear calibration curve was plotted with a correlation coefficient of 0.98.

The precision (repeatability) of the batch adsorption method was studied by measurements of eight replicates of 2.0 mg L^{-1} standard solution of As^{III} as shown in **Table 2**. The precision, expressed in terms of standard deviation (%RSD), was 2.1.

3.5. Application of the avocado seed adsorbent in real water samples

A water sample from East London municipality was adsorbed by the raw avocado seed under the optimized conditions. It is shown in **Figure 6(a** and **b)** that the bio-adsorbent is removed (54 and 55%) from sampling area A and B, respectively. During the adsorption of As^{III} from environmental water samples, an interference can be experienced from metal ions such as Fe^{III}, Fe^{II}, Zn^{II}, Cd^{II}, Ni^{II}, Mn^{II}, Al^{III}, Pb^{II}, and Cu^{II} [10].

This indicated that avocado seed has the great potential in removing heavy metals like As^{III} in environmental water samples without being modified.

Adsorption isotherm	Parameter	Value
Langmuir	K_L	0.0022 L/mg
	R_L	0.17
	R^2	0.97
Freundlich	R^2	0.72

Table 1. Adsorption isotherms for As^{III} adsorption by a bio-adsorbent.

As^{III} standard solution adsorbed by avocado seed	Data
Before adsorption	2.0 mg L^{-1}
After adsorption	(0.50; 0.50; 0.50; 0.50; 0.49; 0.49; 0.49; 0.47) mg L^{-1}
Mean	0.49 mg L^{-1}
Standard deviation	± 0.010 mg L^{-1}
Relative standard deviation	2.1%

Table 2. The repeatability of As^{III} concentration during the adsorption by avocado seed obtained from a local shop in Johannesburg.

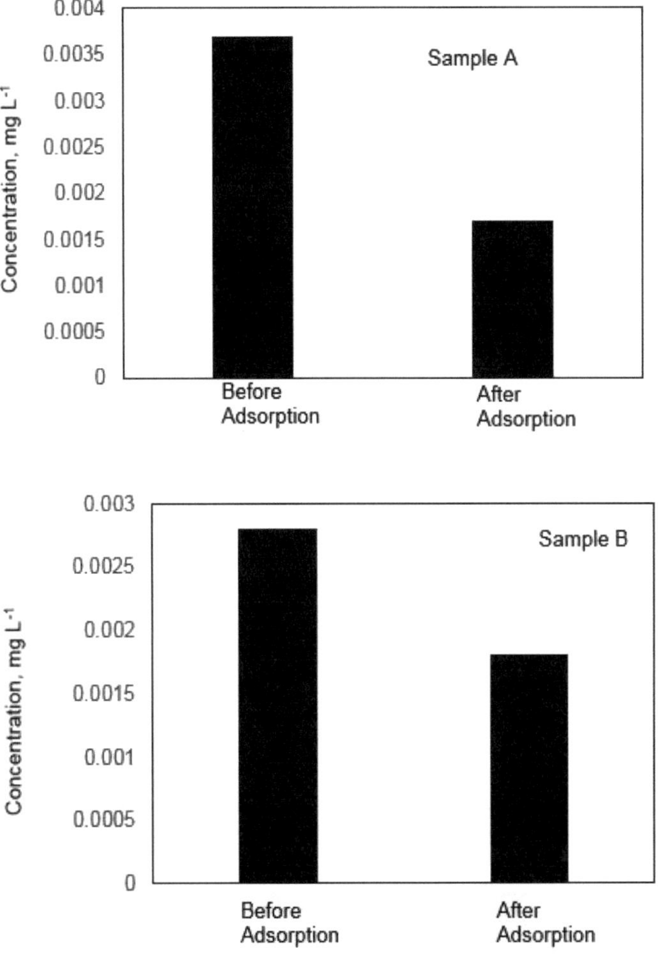

Figure 6. The determination of the percentage (%) of As^{III} removed by the bio-adsorbent from (a) Sample A to (b) Sample B using ICPOES.

4. Conclusions

This work focused on removal of arsenic from aqueous solution using a powdered raw avocado seed. Important parameters that affect adsorption were optimized accordingly, pH 6, analyte 2 mg L^{-1}, dosage mass 0.8 g, and contact time 120 min, and temperature was constant from room temperature of 40°C. It was observed that raw avocado fruit seed can remove more than 50% of arsenic(III) from real water sample without any modification. The advantages of this bio-adsorbent is that it requires less preparation time and is readily available. The use of avocado seeds as a bio-adsorbent will also reduce the waste that is normally discarded in the streets, and it does not affect the food security issues since it is not edible. Due to the advantages that it possesses, it is strongly recommended that it should be incorporated in the removal of toxic heavy metals such as As^{III}. The adsorption isotherm data were tested for both Langmuir and Freundlich models. The regression coefficient and R_L values, best fitted Langmuir model (R^2 = 0.97), and the adsorption capacity was 93.75 mg/g. The Langmuir model means that chemisorption took place at the monolayer of the bio-adsorbent due to the availability of functional groups such as carboxylic acids that have high affinity for metal ions such As^{III}.

Acknowledgements

This work was supported by the National Research Foundation of South Africa (Thuthuka Grant No. 107066 and CPRR Grant No. 98887); the Centre for Nanomaterials Science Research; the National Nanoscience Postgraduate Teaching and Training Platform; the University of Johannesburg (UJ), South Africa; and the Faculty of Science, University of Johannesburg (UJ), South Africa.

Author details

Khathutshelo Catherine Mqehe-Nedzivhe, Khathutshelo Makhado,
Oluwasayo Folasayo Olorundare, Omotayo Ademola Arotiba, Elizabeth Makhatha,
Philiswa Nosizo Nomngongo and Nonhlangabezo Mabuba*

*Address all correspondence to: nmabuba@uj.ac.za

University of Johannesburg, Johannesburg, South Africa

References

[1] James KA, Meliker JR, Nriagu JO. Arsenic. In: Quah SR, Cockerham WC, editors. The International Encyclopedia of Public Health, 2nd ed. vol. 1. Oxford: Academic Press; 2016; **27**:170-175

[2] Bissena M, Frimmel FH. Arsenic—A review. Part II: Oxidation of arsenic and its removal in water treatment. Clean Soil Air Water. 2003;**31**(2):97-107. DOI: 10.1002/aheh.200300485

[3] Nahar N, Mandal A. Phytoremediation of arsenic from the contaminated soil using transgenic tobacco plants expressing ACR2 gene of Arabidopsis thaliana. Journal of Plant Physiology. 2017;**218**:121-126. DOI: 10.1016/j.32 jplph.2017.08.00

[4] Patil-Shinde V, Mulani KB, Donde K, Chanvan NN, Ponrathnam S, Tambe SS. The removal of arsenite [As(III)] and arsenate [As(V)] ions from wastewater using TFA and TAFA resins: Computational intelligence based reaction modeling and optimization. Journal of Environmental Chemical Engineering. 2016;**4**(4):4275-4286. DOI: 10.1016/j.jece.2016.09.030

[5] Çiftçi TL, Henden E. Nickel/nickel boride nanoparticles coated resin: A novel adsorbent for arsenic(III) and arsenic(V) removal. Powder Technology. 2015;**269**:470-480. DOI: 10.1016/j.powtec.2014.09.041

[6] Banerji T, Chaudhari S. Arsenic removal from drinking water by electrocoagulation using iron electrodes—An understanding of the process parameters. Journal of Environmental Chemical Engineering. 2016;**4**(2016):3990-4000. DOI: 10.1016/j.jece.2016.09.007

[7] Vahidnia A, van der Voet GB, de Wolff FA. Arsenic neurotoxicity—A review. Human and Experimental Toxicology. 2007;**26**(10):823-832

[8] Vaezi M, Bahar B, Mousavi A, Yaghmai M, Kasaeian A, Souri M, Jahani M, Alimoghaddam K, Ghavamzadeh A. Comparison of 60 and 80 mg/m^2 of daunorubicin in induction therapy of acute myeloid leukaemia. Heamatological Oncology. 2017;**35**(1):101-105. DOI: 10.1002/hon.2236

[9] Lakovleva E, Maydannik P, Ivanova TV, Sillanpää M, Tang WZ, Mäkila E, Salonen J, Gubal A, Ganeev AA, Kamwilaisak K, Wang S. Modified and unmodified low-cost iron-containing solid wastes as adsorbents for efficient removal of As(III) and As(V) from mine water. Journal of Cleaner Production. 2016;**133**:1095-1104. DOI:10.1016/j.jclepro.2016.05.147

[10] Magoda C, Nomngongo PN, Mabuba N. Magnetic iron–cobalt/silica nanocomposite as adsorbent in micro solid-phase extraction for preconcentration of arsenic in environmental samples. Microchemical Journal. 2016;**128**:242-247. DOI: 10.1016/j.microc.2016.05.005

[11] Mohan D, Pittman CU Jr. Arsenic removal from water/wastewater using adsorbents—A critical review. Journal of Hazardous Materials. 2007;**142**(1–2):1-53. DOI: 10.1016/j.jhazmat.2007.01.006

[12] Ansari R, Sadegh M. Application of activated carbon for removal of arsenic ions from aqueous solutions. E-Journal of Chemistry. 2007;**4**(1):103-108. DOI: 10.1155/2007/829187

[13] Lui Y, Ma S, Chen J. A novel pyro-hydrochar via sequential carbonization of biomass waste: Preparation, characterization and adsorption capacity. Journal of Cleaner Production. 2017;**176**:187-195. DOI: 10.1016/j.jclepro.2017.12.090

[14] Moyo M, Chikazaza L, Dekhil AB, Hannachi Y, Ghorbel A, Boubaker T. Bioremediation of lead(II) from polluted wastewaters employing sulphuric acid treated maize tassel biomass. Journal of Chemistry and Ecology. 2013;**5**:689-695. DOI: 10.4236/ajac.2013.412083

[15] Pérez-Sirvent C et al. Geochemical background levels of zinc, cadmium and mercury in anthropically influenced soils located in a semi-arid zone (SE, Spain). Geoderma. 2009; **148**:307-317. DOI: 10.1016/j.geoderma.2008.10.017

[16] Olorundare O. Potential application of activated carbon from maize tassel for the removal of heavy metals in water. Physics and Chemistry of the Earth Parts A/B/C. 2012;**50**:104-110. DOI: 10.1016/j.pce.2012.06

[17] El-Nabarawy T, Petro NS, Abdel-Aziz S. Adsorption characteristics of coal-based activated carbons. II. Adsorption of water vapour, pyridine and benzene. Adsorption Science & Technology. 1997;**15**(1):47-57. DOI: 10.1177/026361749701500105

[18] El-Nabarawy T, Petro NS, Abdel-Aziz S. Adsorption characteristics of coal-based activated carbons. 1. Adsorption of nitrogen and carbon dioxide. Adsorption Science & Technology. 1996;**13**(3):177-186. DOI: 10.1177/026361749601300304

[19] Olorundare O. Activated carbon from lignocellulosic waste residues: Effect of activating agent on porosity characteristics and use as adsorbents for organic species. Water, Air, & Soil Pollution. 2014;**225**(3):1-14

[20] Olorundare O. Steam activation, characterisation and adsorption studies of activated carbon from maize tassels. Chemistry and Ecology. 2014;**30**(5):473-490

[21] Bhaumik M. Highly effective removal of toxic Cr(VI) from wastewater using sulfuric acid-modified avocado seed. Industrial & Engineering Chemistry Research. 2014;**53**(3):1214-1224. DOI: 10.1021/ie402627d

[22] Díaz-Muñoz LL. Sorption of heavy metal ions from aqueous solution using acid-treated avocado kernel seeds and its FTIR spectroscopy characterization. Journal of Molecular Liquids. 2016;**215**:555-564. DOI: 10.1016/j.molliq.2016.01.022

[23] Zhang S. Arsenite and arsenate adsorption on coprecipitated bimetal oxide magnetic nanomaterials: $MnFe_2O_4$ and $CoFe_2O_4$. Chemical Engineering Journal. 2010;**158**(3):599-607. DOI: 10.1016/j.cej.2010.02.013

[24] Mandal S, Padhi T, Patel R. Studies on the removal of arsenic(III) from water by a novel hybrid material. Journal of Hazardous Materials. 2011;**192**(2):899-908. DOI: 10.1016/jhazmat.2011.05.099

[25] Ren Z, Zhang G, Chen JP. Adsorptive removal of arsenic from water by an iron–zirconium binary oxide adsorbent. Journal of Colloid and Interface Science. 2011;**358**(1):230-237. DOI: 10.1016//j.jcis.2011.01.013

[26] Zhu Y. Avocado seed-derived activated carbon for mitigation of aqueous ammonium. Industrial Crops and Products. 2016;**92**:34-41. DOI: 10.1016/j.indcrop.2016.07.016

[27] Desta MB. Batch sorption experiments: Langmuir and Freundlich isotherm studies for the adsorption of textile metal ions onto teff straw (*Eragrostis tef*) agricultural waste. Journal of Thermodynamics. 2013;**2013**(2013):1-6. DOI: 10.1155/2013/375830

[28] Hall KR, Eagleton LC, Acrivos A, Vermeulan T. Pore and solid kinetics in fixed bed adsorption under constant pattern conditions. Industrial and Engineering Chemistry Fundamentals. 1966;**5**(2):212-223. DOI: 10.1021/i160018a011

Section 2

Health Risks

Chapter 4

Food and Water Security as Determinants of the Mitigation of Health Risks Due to Exposure to Arsenic in Water

Rebeca Monroy-Torres

Additional information is available at the end of the chapter

http://dx.doi.org/10.5772/intechopen.76977

Abstract

Exposure to arsenic is a global public health problem, and the effects on health are several from cancer to metabolic diseases such as diabetes and hypertension. The metabolism and excretion depends on having a good nutritional status and the latter of an adequate diet. It is known that the consumption of certain trace elements and nutrients intervene in the metabolism, in the excretion and in the protection of the adverse effects that the metalloid has on the organism. The amount of proteins consumed, the type of amino acids such as cysteine, methionine; vitamins such as C, thiamin, vitamin B_{12}, folic acid, minerals such as calcium and other nutrients such as fiber have been studied and associated with a lower concentration of As in blood and urine, as well as minor dermatological lesions as well as other organs and systems. A study by Monroy-Torres et al. (2018, *in press*), in adolescents exposed to As in water, found greater excretion of As with a 4-week vitamin supplementation, increasing iron levels, after the intervention. Reason for which this chapter, shows a review of the main evidence of health impact indicators that can lead to mitigate the effects of exposure to As across to promote food security, access to cleaner drinking water and good nutrition.

Keywords: food security, water security, nutrients, vitamins, nutritional status

1. Introduction

1.1. Food and water security: an era of sustainable development

With the demographic growth, the lack of a culture of water care and the overexploitation of aquifers, the problems of higher levels of arsenic in water for human consumption have

© 2018 The Author(s). Licensee IntechOpen. This chapter is distributed under the terms of the Creative Commons Attribution License (http://creativecommons.org/licenses/by/3.0), which permits unrestricted use, distribution, and reproduction in any medium, provided the original work is properly cited.

increased, and if there is no solution, the problem will be exacerbated and with its impacts on health. Before continuing with the main approach of this chapter, I will integrate an issue that should be mentioned: lack of food security in the people. The food security is the cornerstone for mitigating the effects that arsenic on people's health and to consider the era of sustainable development.

The Sustainable Development Goals (SDGs) (2015–2013) have been established, leaving behind the millennium development goals, derived from ever more prevailing needs and facing the challenges of the health complexities faced by the population. These are derived from investigations that generate solutions to the great problems of humanity, such as achieving the right to health. Health is a human right, but for this to be achieved, governments must establish the conditions of access to basic services and one of them is access to water and food, among others. The SDGs, which began in 2015, were promoted at the United Nations Conference on Sustainable Development, held in Rio de Janeiro in 2012, with the purpose of creating a set of global objectives related to global environmental, economical, political, and social challenges [1].

The SDG are the pillars for the countries promote in their agenda, actions, and strategies to end poverty, to act for the benefit of the environment and the planet, and ensure that all people enjoy peace and prosperity. The 17 SDGs begin with the objective "Eradicating poverty is the pillar and basis of the 2030 Agenda," which is, therefore, the goal number 1 [2]. All the SDGs are interrelated and are conducive to the complexity that is experienced, such as having solutions for the fight against poverty. More than 800 million people still live in poverty, with only 1.25 dollars a day; they have food shortages, as well as water and public services. This increase will generate the need to obtain an income through a work to obtain food, water, health services, and housing [1].Goal #6 corresponds to "Clean water and sanitation" [3], where more than 40% of the population does not have access to potable water, and it is estimated that by 2050, one in four people will not have access to water, which impacts in infectious and diarrheal diseases; and finally, goal #3 "health and well-being" [4] would be integrated, where the figures for diseases derived from the lack of drinking water, food safety, and vaccination, require further research. The line of research that leads the name "*Environmental Nutrition and Food Security,*" whose origin and foundation can be reviewed [5].

The line integrates a model with the interdisciplinary approach and integrating topics that have been fragmented over time, to understand the conditions of environmental impacts such as food safety and water safety in people. Food security is achieved when people always have constant access to adequate food, without creating health risks, and being constant and permanent. It is studied from the pillars of stability, availability, access, and biological incorporation [6]. In this case, studying food safety without considering water safety, would be walking and moving blindly or as if this complexity could be analyzed and understood from a single methodological paradigm or study design type.

The issue of water safety, would carry the following indicators for the 17 sustainable development objectives (**Table 1**) [1], based on the hypothesis that if a person has a good nutritional status derived from food security, environmental stressors (water contaminated with lead, arsenic, for example), respiratory exposures, viruses or bacteria in food would have less

SDGs	Example of the central theme "Arsenic in drinking water in the population"
No poverty	Access to public services and water sanitation as well as healthy, local and fresh foods.
Cero Hunger	Access to drinking water permanently for food preparation.
Good health and well-being	Access to drinking water for the prevention of foodborne diseases, infectious diseases and diarrheal diseases.
Quality education	Teaching techniques for the purification of water, the culture of water, as well as occupational risks, to achieve a prevention education.
Gender equality	Opportunity for all, considering that women provide water and food in most homes in the world.
Clean water and sanitation	Compliance with the law, through the monitoring and supervision of water quality in homes
Affordable and clean energy	Harvest of rainwater, use of solar energy to heat water.
Decent work and economic growth	Occupational health and risk reduction in work areas. Decent Salary.
Industry, innovation and infrastructure	Sustainable production programs, water treatment plants, rainwater harvesting in companies
Reduced inequalities	Opportunities for all, with quality education, gender inclusion and access to public services.
Sustainable cities and communities	Construction of recreational spaces, social security. Design and promotion of green spaces
Responsable consumption and production	Green industry, use of additives and ingredients that do not cause harm to the consumer.
Climate action	Sustainable feeding, care of water and environment.
Life below water	Care and protection of marine life.
Life of land	Protection of ecosystems and protected areas
Peace, justice and strong institutions	Actions to reduce violence.
Partnerships for the goals	Institutional links and synergies in actions that lead to address social problems in common, with less cost, inputs and time.

With permission: Monroy-Torres R.

Table 1. A proposal for an agenda in research based on the 17 SDGs with the research line of "Environmental Nutrition Food Safety" and the topic water.

impact or manifestations (less risk of developing or manifest an outcome) than populations or people with a poor nutritional status (obesity, low weight, etc.) [7].

Given this introduction and justification, I share the studies we have done with my research team since 2005 in the State of Guanajuato, Mexico. The difficulties to achieve impact even with the scientific evidence, as well as the proposals and strategies that we have had to raise for this constant search to solve problems with the investigations carried out. While there are many pollutants as conditioning factors to which people are currently exposed, the actions and approach of arsenic from the approach of environmental nutrition and food security, and water can be triggers for a new and different way of decrease the damage to the health with the different pollutants:

> *"Time to give focus to public health and with it the promotion of food security as a strategy to mitigate the effects and risks to health"*

1.2. Arsenic: food and water security

Arsenic is a metalloid that can be found naturally in the Earth, mainly in the Earth's crust, and can leak into the groundwater reserves. In many countries of the world, groundwater represents the main source of drinking water; therefore, exposure to arsenic from drinking water is considered a public health problem [8, 9]; not only the direct consumption of water contaminated with arsenic is a factor of exposure but also indirect consumption through food, when water is used in the preparation of food, in the irrigation of fields, and for consumption of animal, is an exposure factor [10].

The World Health Organization (WHO) and the guidelines of the Environmental Protection Agency (EPA) of the United States, established an allowed limit of 0.01 mg/L for arsenic in drinking water [11]. The Ministry of Health of Mexico establishes an allowable limit of 0.05 mg/L of arsenic in drinking water [11, 12], with current modifications at 0.025 mg/L, which are still above international standards.

1.3. Metabolism

To study and know the toxicity of the metalloid, must be made of the speciation of it, but also of the physiological conditions of the exposed population (age, body surface, nutritional status, etc.), which will be addressed in this chapter. Arsenic compounds are classified into three groups: (1) The inorganic arsenic compounds; (2) Organic compounds, and (3) Gaseous compounds; they are usually in their trivalent and pentavalent state. The most common trivalent inorganic arsenic compounds are arsenic trioxide, arsenic trichloride, and sodium arsenic; the most common pentavalent inorganic arsenic compounds are arsenic pentoxide, arsenic acid, and arsenate; and the organic arsenic compounds are monomethylarsonic acid (MMA) and dimethylarsinic acid (DMA) [13, 14]. The level of toxicity of arsenic depends on its valence and its organic or inorganic form. Organic arsenic is considered less toxic than inorganic as it is easier to excrete. The inorganic arsenic is the most toxic form of As, in its trivalent form it can combine chemically with the sulfhydryl groups, which are organic compounds that contain a sulfur atom as a functional group attached to a hydrogen (-SH), these functional groups form inter- and intramolecular bridges in proteins and their structure and biological function depend on them.

1.4. Absorption

Arsenic can be incorporated into the body by ingestion, inhalation or through the skin, more than 90% of the ingested arsenic is absorbed in the gastrointestinal tract. Organic arsenic is easier to excrete and is usually found mainly in meat, seafood, and some cereals. The ability of trivalent arsenic (AsIII) to join groups -SH, confers that toxic capacity while pentavalent arsenic (AsV) interferes with phosphorylation reactions, due to its similarity chemical with phosphate [13]. In food, the organic form is the most frequent and is most susceptible to elimination, except for some types of fish, crustaceans, and algae, in most foods (fruits, vegetables, cereals, meats, etc.) that have arsenic concentration lower than 0.25 mg/kg [13]. After ingestion, arsenic is absorbed in the gastrointestinal tract, then passes into the portal circulation and reaches the liver.

The metabolism of arsenic depends on various reduction and methylation reactions in which enzymes and compounds such as s-adenosylmethionine (SAM), arsenic methyltransferase (AS3MT), glutathione (GSH), and oxidized glutathione (GSSG) participate. Initially, GSH participates in the reduction of AsV to AsIII, which is pH dependent and is influenced by the presence of other substances that can be reduced or oxidized; subsequently, the AS3MT participates in the transfer of a methyl group of the SAM enzyme to the trivalent inorganic arsenic to generate MMA(V) Ácido monometilarsónico (monomethylarsonic acid), this compound is reduced to MMA(III) Ácido metilarsenioso (monomethylarsonous acid) by the action of a specific reductase and in this reaction participate the AS3MT, the GSH, and the GSSH, between other compounds; MMAIII enters a methylation reaction to generate DMA(V) Ácido dimetilarsínico (dimethylarsinic acid), and finally, DMAV can be reduced to DMA(III) Ácido dimetilarsenioso (dimethylarsinous acid). These reactions are carried out regardless of the route of absorption [13, 15].

1.5. Excretion

The excretion of As is variable among people; it will depend on the time and dose of exposure to As, as well as the efficiency of the methylation reactions [8]. It is known that the ability to meet arsenic is saturated when the ingestion exceeds 0.5 mg/d [16]. The main metabolites found in urine after acute or chronic exposure to arsenic are inorganic arsenic, MMA and DMA [8, 17], residual As can also be found in keratinized tissues (skin, hair, and nails), where the presence of arsenic in these tissues has a significant association with chronic exposure to metalloid [15]. Although in general, the presence of arsenic in urine is considered an indicator of acute exposure, it has been found that in prolonged exposures arsenic levels in urine are maintained and increase [8].

That is why the presence of arsenic in urine is considered a reliable marker of chronic exposure to As. It is known that the usual consumption of arsenic varies between 5 and 25 μg/d, and the excretion of arsenic in urine is usually less than 25 μg in 24-hour urine. It has been reported that after a consumption of seafood products, the concentration of arsenic in urine can increase to 300 μg in 24 hours, however, this concentration will decrease after 1 day to <25 μg in 24-hour urine [18]. An excretion rate greater than 1000 μg in 24-hour urine is a sign of significant exposure to As [19, 20].

1.6. Effects of acute and chronic exposure: signs and symptoms

Chronic exposure to inorganic arsenic causes, mainly, skin lesions, cancer of the bladder, kidney, lungs, and liver. Keratosis and changes in skin pigmentation are recognized signs of chronic exposure to arsenic, while melanosis is a sign associated with acute exposure [8, 21]. Cumulative doses in adults of 0.10 mg/kg/d are related to signs of chronic intoxication and usually appear after 2–8 weeks of exposure. It also causes alterations in intestinal epithelial tissue, decreased production of red and white blood cells, central nervous system involvement, liver damage, and kidney damage [17, 22–24]. The acute lethal dose of arsenic in humans has been estimated at 0.6 mg/kg/day [25].

Exposure to arsenic generates oxidative stress, and this represents the main mechanism by which arsenic generates multi-organ damage, there are free radical formation directly and also inhibitory effects on antioxidant components [23]. An alteration in the metabolism of GSH secondary to the union of arsenic with -SH groups has been found, as well as the elevation of genes associated with oxidative stress such as heme oxygenase-1 and metallothioneins, after chronic exposure to arsenic [17, 26]. Chronic exposure to arsenic causes an increase in the production of reactive oxygen species (ROS) and an alteration in defense mechanisms against the pro-oxidative effect of ROS, mainly the catalase enzyme (CAT), superoxide dismutase (SOD), and glutathione peroxidase (GSH-Px) [23].

1.7. Role of nutrition: food security

The effect of several nutrients that can reduce arsenic toxicity has been studied. Oral supplementation with cysteine, methionine, vitamin C, and thiamin at a dose of 25 mg/kg in studies in mice that were exposed to arsenic (10 ppm of sodium arsenite in drinking water) was observed a decrease in oxidative stress caused by exposure to arsenic, these nutrients were able to produce specific changes in the levels of lipid peroxidase, antioxidant enzyme activity and reduction in the concentration of arsenic in blood between 3 and 11%, in liver of 26–37%, and in kidneys from 16 to 24% ($p < 0.05$). Methionine and cysteine are the main amino acids that participate in the metabolism of arsenic, and vitamin C is an antioxidant [23, 27].

A case–control study conducted by Mitra et al. [28] in the adult population ($n = 265$) exposed to arsenic in drinking water (<500 μg/L) belonging to the city of West Bengal in India, found a higher probability of lesions due to exposure to arsenic associated with low protein intake (odds ratio = 1.94; CI: 95%, 1.05–3.59), calcium (OR = 1.89, 95% CI, 1.04–3.43), fiber (OR = 2.02, 95% CI, 1.15–4.21), and folic acid (OR = 1.67 95% CI, 0.87–3.2). In a study involving an adult population of 18–65 years ($n = 1016$), exposed to arsenic in drinking water in the city of Bangladesh, the consumption of some nutrients was quantified, finding that the ingestion of cysteine (950–2005 mg/d), methionine (1745–3530 mg/d), calcium (407–1500 mg/d), proteins (77–141 g/d), and vitamin B12 (2.3–9.9 μg/d) were associated with minors percentages of inorganic arsenic in urine, this speaks of a decrease in the total concentration of inorganic arsenic. Similarly, high ingestion of niacin and choline were associated with a higher rate of AMD to MMA ($p = 0.02$ for both cases), which indicates a higher conversion rate, and therefore, greater excretion of it [29]. A study developed in adult population (38 ± 10 years) ($n = 130$), evaluated the effect of folic acid on blood arsenic levels, the intervention consisted of an oral supplement of 400 μg/day of folic acid for 12 weeks. The authors found a decrease in the concentration of MMA in blood, going from 22.24 ± 2.86% in the intervention group and 1.24 ± 3.59% in the control group ($p < 0.0001$). Total arsenic in blood was reduced by 13.62% in the intervention group, and 2.49% in the control group ($p = 0.0199$), as well as an increase in the rate of DMA excretion ($p = 0.099$), folic acid participates in methylation reactions [30, 31].

Zablotska et al. [32] studied 11,746 people over 18 years of age and found that consumption of riboflavin, pyridoxine, vitamin A, C, and E, and folic acid significantly modified the effects of arsenic exposure. Participants in the highest percentile of consumption of each nutrient (≥1.17 mg/d of riboflavin, ≥4.19 mg/d of pyridoxine, ≥351.61 μg/d of folic acid, ≥7113.06 mg/d of vitamin C, and ≥ 6.41 mg/d of vitamin E), showed a reduction in the risk of lesions derived from exposure to arsenic in 46–68% ($p < 0.05$). This reflects that the consumption of these key nutrients not only increase the excretion of arsenic, but also the reduction of adverse effects derived from exposure to As. In addition, it has been observed that the concentration of selenium in blood, which is an essential trace element required for the synthesis of several proteins, is inversely related to the risk of pre-malignant lesions in the skin, finding an inversely proportional relationship ($p = 0.03$) between the concentration of selenium in blood and arsenic in urine [8].

In addition to the above, a poor nutritional status, mainly malnutrition, correlates with the development of skin lesions caused by arsenic poisoning. In 2007 Maharjan et al. [33] developed a study in the adult population ($n = 539$) where an increase in the risk of manifesting skin lesions due to the arsenic exposure was found 1.65 times more in subjects with a body mass index (BMI) per below normal (16.5–17.1 kg/m^2) compared to those with a normal BMI ($p < 0.05$). In addition, it has been described that the decrease in BMI is also a non-specific manifestation of chronic exposure to high concentrations of arsenic (daily arsenic intake of 30 μg/kg/d). In rural populations, a low BMI is a reflection of a poor nutritional status, which is associated with a low intake of certain nutrients, including antioxidants, and nutrients whose poor ingestion has been related to increased production of MMA, the toxic form of arsenic, and a low production of DMA [30, 31, 34].

1.8. Evidence in Guanajuato, Mexico: food and water security

During 2002–2003, the State Public Health Laboratory of Guanajuato found, in two communities of the state, San Agustín of the municipality of Irapuato and Cútaro in Acámbaro, amounts of arsenic in well water (0.950 and 0.109 mg/L, respectively), greater than the permissible limits established by the WHO, and the Secretary of Health of Mexico. These communities have been studied for several years, adding other populations, where this metalloid has also been found.

Next, in **Table 2**, a chronology of the studies published and in the process of publication is presented. The first studies were based on identifying a relationship between the consumption of water contaminated with arsenic and the presence of arsenic in the hair of children living in two communities, which had a relationship ($n = 55$, $p < 0.0001$) [35]. In addition, a survey of mothers of children of the previous study revealed that 90–94% of them use well water for different culinary preparations such as broths, soups or beans, even with the knowledge that this source of water is contaminated with arsenic, according to the participants, this is done for lack of economic resources to acquire drinking water [10].

References	Main evidence
Effect of four-week multivitamin supplementation on nutritional status and urinary excretion of arsenic in adolescents *Rebeca Monroy-Torres, Espinoza Pérez JA, Ramirez Gomez X, Carrizalez Yañez L, Linares-Segovia B, Mejía Saavedra JJ* *Aceptado 2018, en la revista de Nutrición Hospitalaria*	Our objective was to measure the effect of multivitamin supplementation on the nutritional status and urinary excretion of arsenic in adolescents exposed to this metal through drinking water. With an intervention study was carried out on 45 adolescents, exposed to arsenic in drinking water, who were given a daily multivitamin supplement for 4 weeks. The nutritional status, and the levels of arsenic in urine and drinking water were evaluated weekly. The main results were Basal nutritional intake was low for protein, fiber, folic acid, vitamin B2, B6, B12, E, C, selenium and iron, increasing its consumption through the supplement during the intervention and with an increase of approximately 1 g/dL of hemoglobin in all participants. At the end of the intervention the participants presented increase of fat-free mass and decrease in the percentage of body fat. The urinary excretion of arsenic, was greater [35.91 μg/gCr (95% CI = 23.2–74.8 μg/gCr)] in the first week of intervention ($p < 0.05$) compared to baseline levels of urinary arsenic [43.2 μg/gCr (95% CI = 30.8–117.6 μg/gCr)] and an average of As in water of 96.2 ± 7.5 μg/L. Four-week multivitamin supplementation in the adolescent population studied improved nutritional status and increased metalloid excretion significantly in the first and second post-intervention weeks.
Experiences around the lack of access to water in homes in the State of Guanajuato, Mexico *Rebeca Monroy Torres, Jaime Naves Sánchez, Hugo Melgar-Quiñonez* (Enviado a publicación a la revista Española de Nutrición Comunitaria)	We analyze the experience in households in the State of Guanajuato, Mexico that suffer from limitations regarding access to water in quality and quantity. A survey of 17 items was applied to 352 households (female heads of household) to measure experiences regarding access to water, in addition to food security, schooling and sociodemographic aspects. Where 33.4% of households reported concern about not having access to water and 74.8% did not have access. 70.8% had to buy water to drink and 5.7% got sick and related it to water consumption. 65.6% of households showed food insecurity. The correlation was significant for the level of education of female heads of household, households with children under 1 and 12 years old with the use of tap water, preparing powdered milk for children, and for food at home and water. These experiences of households around access to water contribute to the discussion and development of scales that insecurity to water, considering food security.
Monroy-Torres y Espinoza-Pérez [39]	The objective of the present work was to identify the presence of risk factors of exposure to arsenic contamination in water, in population living in areas where high levels of this metalloid have been detected. For the identification of the risk factors that could intensify the metalloid exposure, an analytical and transversal study was carried out. For the measurement of food security, a scale validated for Latin America and the Caribbean was applied to 30 heads of family responsible for food, in addition to 30 items, with a three-month time limitation that measured the main risk factors of exposition, which were integrated from the derived risk factors selected from previous studies in the communities exposed to the metalloid and from the scientific literature collected so far. The main risk factors were: food insecurity in 70% of households, 63% worried about not having access to water, lack of access and availability of drinking water and 4% did not have access to water during the last 3 months, since 40% used well, pipe or tap water without previous treatment to prepare food, and 83.3% used to prepare beverages; Other risk factors to consider were the education level of the head of the family and access to public services. The identification of the main risk factors will allow the design of a validation scale, for screening and preventing possible arsenic poisoning in communities exposure to arsenic in water in addition to measuring food security.
Monroy-Torres R et al. [38]	The objective of the study was to evaluate feeding and nutrition practices in communities of the state of Guanajuato exposed to arsenic and to identify some indicators of nutritional risk that contribute to the health effects of the metal. With a transversal design, a survey was applied to 30 family heads, who were selected from a previous study, culinary practices, food consumption, sociodemographic characteristics were evaluated. Culinary and food practices were detected as risk indicators in a population exposed to arsenic. Therefore, these practices should be considered as indicators in the evaluation of the health effects of exposure to the metalloid and other pollutants.

References	Main evidence
Monroy-Torres y cols [37]	To identify the state of food security and access to water in Mexican households, using a validated scale applied in 352 households in rural and urban communities, as well as a pilot scale to assess access to water. Where a 73% of households were classified as food secure (level 1), 15% as being mildly food insecure (level 2), 7% were considered food insecure at a moderate level (level 3) and only 4% were households with severe food insecurity (level 4). At all food security levels interviewees were worried about not having enough access to water, and in the last 3 months 50% of households experiences water scarcity. Most of the households used tap water to prepare the milk for the children, as well as for personal hygiene. 25% of the interviews reported that water availability has declined in their households, while the cost of water has increased. Food security cannot be conceived without taking into account the water situation. The majority of the households report lack of access to enough water, which usually does not meet the conditions of safety for human consumption and food preparation.
Monroy-Torres R et al. [36]	The objective of this study was to describe the perception of the beneficiaries of the Oportunidades program, about food and nutrition security in Atarjea, Guanajuato. A descriptive and qualitative approach study was carried out in 10 families, with a population under 10 years old, in two Atarjea localities. A nutritional and validated food safety survey was applied. Families and doctors do not perceive that the program is fair. New and constant evaluations are suggested to the program by external personnel and by the same beneficiaries of the program. On a proportional basis, families presented food insecurity with moderate hunger and mild food insecurity.
Monroy Torres R et al. [10]	The chronic exposition to arsenic has been linked to various health problems, and some studies have been winged as a source of exposition, in addition to water and food. In some communities in the state of Guanajuato levels were detected outside the rule of arsenic in drinking water. OBJECTIVE: To describe the accessibility to safe water for consumption and preparation food in a community exposed to groundwater with arsenic. MATERIAL Y METHODS: Survey conducted across in55 housewives from 27 to 55 years of age. The questions were about the use of well water in the preparation in food such as broths, soups, beans, water, fruit; ingestion milk background; main crops in the community, raising cat le and source of drinking water for them. RESULTS: The 90% of housewives used well water for drinking and preparing their food. With regard to the consumption of the well of milk, 24 (44%) consumed milk brand and 31 (44%) consumed cow's milk. The well water was used for livestock and to irrigate crops. CONCLUSIONS: It is emerging that seeks strategies for the population for to have access to water safe, and that health risks are predictable with these sources of exposition to arsenic, like cancer, diabetes and hypertension.
Monroy-Torres R et al. [35]	This cross-sectional study measures the arsenic level in school children exposed to contaminated well water in a rural area in México. Arsenic was measured in hair by hydride generation atomic absorption spectrophotometry. Overall, 110 children were included (average 10 years-old). Among 55 exposed children, mean arsenic level on hair was 1.3 mg/kg (range < 0.006–5.9). All unexposed children had undetectable arsenic levels. The high level of arsenic in water was associated to the level in hair. However, exposed children drank less well water at school or at home tan unexposed children, suggesting that the use of contaminated water to cook beans, broths or soups may be the source of arsenic exposure.

Table 2. Compilation of main evidence of the conditioning factors in environmental nutrition and food security of 2009–2018, in Guanajuato, México.

Another study, derived from being able to identify other factors of exposure to As, even though 11% drank tap water and 24.4% both tap water and tap water, even though the population was alerted about concentrations of the metalloid, most reported that in their homes tap water was used to prepare food under cooking as well as in the preparation of flavor water or fruit.

The use of direct water from the tap was treated with drops of chlorine or silver nitrate, which reflects that the population has knowledge about the microbiological form of disinfection, but not for the removal of metals or other toxic substances. The trust that people perceived that they used tap water for food, because they use to boiling the water became drinkable and safe. But it is known to boil water from food or with food, only eliminate microbiological and non toxicological risks, such as eliminating arsenic in water [10, 35].

The first evidence in two communities of the state of Guanajuato (Cútaro and San Agustín) of the presence of arsenic in water and food consumption, biological incorporation to metal, together with environmental factors such as feeding and nutrition practices, socioeconomic, deficiency in public services and lack of access to an innocuous source of water, reflect a risk to the health of the inhabitants of these communities. On the other hand, the evidence shows that the greater the deterioration of the nutritional status, the greater the harmful effects in the organism derived from exposure to arsenic. In addition, several studies show that the consumption of several nutrients, in isolation, improves the detoxification of the metal and reduces the damage to the organism [36–39].

The poor nutritional status in these populations, either due to the excess or nutritional deficiency, the history that supplementation with some nutrients improves the nutritional status, and therefore, the metabolism of the metal in the body, besides that there is no evidence of treatment with A multivitamin supplementation scheme in school population exposed to arsenic are factors that justify the analysis of a multivitamin supplementation plus a dietary regime, on the urinary excretion of arsenic in the school population of the communities of San Agustín and Cútaro as well as others that there are high levels of the metal [40].

The theme of the study of food and water security, derived from the experience of studies in the population of Guanajuato, as intervention proposals from the nutritional clinic, derived from the basis of its metabolism (absorption and excretion) where the consumption of nutrients and access to nutrients is through food and these when people have food security [6] and food security achieved an adequate nutritional status and thus a way to mitigate the toxicity problems of the metalloid. Reason for which the designs of the studies integrated the dietary evaluation and the nutritional status until to design a clinical nutritional intervention study in a follow-up population, entitled "*Effect of a multivitamin supplementation of 4 weeks on the nutritional status and urinary excretion of arsenic in adolescents*." The main objective was to measure the effect of multivitamin supplementation on nutritional status and urinary excretion of arsenic in 45 adolescents exposed to the metalloid through drinking water. The vitamin supplement was given daily, for 4 weeks, and to the adolescents. Nutritional status, arsenic levels in urine and drinking water were measured too. The main findings were a low intake for proteins, fiber, folic acid, vitamin B_2, B_6, B_{12}, E, C, selenium, and iron, increasing their consumption through the supplement during the intervention, and with an increase of approximately 1 g/dL of hemoglobin, increase in fat-free mass, decrease in the percentage of body fat, with a higher urinary excretion of arsenic [35.91 μg/g Cr (95% CI = 23.2–74.8 μg/g Cr)] from the 1st week of intervention, which was statistically significant in comparison with basal levels of urinary arsenic [43.2 μg/g Cr (IC95% = 30.8–117.6 μg/g Cr) ($p < 0.05$) with an average water with As of 96.2 ± 7.5 μg/L.

Other data of interest of the study were a higher percentage of body fat in women (women 27.4 vs. 17.2% in men), where it is known that the presence of obesity generates an inflammatory process [41]. A high BMI has been associated with alterations in the metabolism and excretion of As, due to the relationship between sex hormones, as main donor of methyl groups, in addition to SAM, is the choline, its secretion is influenced by the presence of estrogen. Therefore, a higher risk of toxicity in men could be expected, although it is also known that adolescents have better methylation than the adult population [23, 42, 43]. The consumption of vitamin B_6 and folic acid were low, they are essential nutrients for the formation of methionine, which in turn is required in the metabolism of As and in the formation of the SAM cofactor, part of the arsenic methylation process; and on the other hand nutrients such as vitamin C and vitamin E, are recognized antioxidants in reducing oxidative damage caused by arsenic. The nutritional contribution improved with the supplementation, for the case of hemoglobin it increased almost 1 gram during the 4 weeks of treatment [42, 44] as well as polyphenols present in some of the components of the multivitamin (broccoli extract, cranberry, etc.), although they did not have an important effect in the excretion of arsenic, can provide protection to the organism from the oxidative stress that is being produced by exposure to arsenic in these adolescents [44]. In relation to fiber, a deficiency in its consumption is associated with a higher probability of the appearance of dermatological lesions since fiber could decrease the absorption of arsenic in the gastrointestinal tract [30, 44, 45]. Soluble fiber (fructo-oligosaccharides), acts as a prebiotic, and therefore, to a better metabolism of As [44, 45]. Many of the foods consumed traditionally in Mexico are rich in fructo-oligosaccharides, such as beans and other legumes [38]. The consumption of fruits and vegetables has been low among the young population as well as greater food insecurity in Guanajuato households (71.2% of food insecurity), which represents a low consumption of antioxidants that have an important role in As metabolism [46].

This is the first study, which addressed the analysis of the effect of a 4-week vitamin and mineral supplementation, integrating dietary and nutritional variables. Therefore, the interaction of diet and the environment should be studied, as well as the integration of problems such as overweight, obesity, and anemia mainly.

2. Conclusion

Adequate nutrition in childhood and adolescence as well as in all stages of life, should not only be promoted to prevent chronic degenerative diseases, which entails an inflammatory problem, it is important to equip people with the mechanisms that promote the greatest excretion of the metalloid combined with adequate growth and development. There are environmental factors that cannot be avoided, such as exposure to heavy metals through natural sources such as water, but food and water security as determinants of the mitigation of health risks from exposure to arsenic.

Acknowledgements

The author is grateful to the Campus Leon of the University of Guanajuato, for the financial support received for the Department of Research Support and Postgraduate within the "Call 2018, of complementary support for publications of students and professors"; as well as the University Observatory of Food and Nutritional Security of the State of Guanajuato (OUSANEG).

Conflict of interest

None.

Author details

Rebeca Monroy-Torres

Address all correspondence to: rmonroy79@gmail.com

Laboratory of Environmental Nutrition and Food Security, University of Guanajuato, Campus Leon, Mexico

References

[1] ONU. Division for Sustainable Development. 2017. Available from: https://sustainabledevelopment.un.org/about [Accessed: January 15, 2018]

[2] PNUD. No poverty: Sustainable development goals. Available from: http://www.undp.org/content/undp/es/home/sustainable-development-goals/goal-1-no-poverty.html [Accessed: January 11, 2018]

[3] PNUD. Water and sanitation: Sustainable development goals. Available from: http://www.undp.org/content/undp/es/home/sustainable-development-goals/goal-6-clean-water-and-sanitation.html [Accessed: January 11, 2018]

[4] PNUD. Good health and well being: Sustainable development goals. Available from: http://www.undp.org/content/undp/es/home/sustainable-development-goals/goal-3-good-health-and-well-being.html [Accessed: January 11, 2018]

[5] Monroy-Torres R. Environmental nutrition and food safety: A laboratory and an integrated research line for an approach to nutritional problems. REDICINAySA. 2016;**5**(5):16-22. Available from: http://www.ugto.mx/redicinaysa/images/publicaciones/2016/sep-oct-2016V1.pdf [Accessed: January 13, 2018]

[6] FAO. Seguridad alimentaria. Cumbre Mundial sobre la Alimentación. 1996. Available from: ftp://ftp.fao.org/es/ESA/policybriefs/pb_02_es.pdf [Accessed: January 11, 2018]

[7] Monroy-Torres R. Design of research projects in health in a sustainable era. In: Pearson, editor. Mexico. 2018. In print

[8] Chen Y, Parvez F, Gamble M, Islam T, Ahmed A, Argos M, et al. Arsenic exposure at low-to-moderate levels and skin lesions, arsenic metabolism, neurological functions, and biomarkers for respiratory and cardiovascular diseases: Review of recent findings from the Health Effects of Arsenic Longitudinal Study (HEALS) in Bangladesh. Toxicology and Applied Pharmacology. 2009;**239**(2):184-192. DOI: 10.1016/j.taap.2009.01.010

[9] Monroy-Torres R, Rodríguez ME, Ramírez GX. El arsénico y su impacto en la salud humana. Naturaleza. Universidad de Guanajuato. 2010;**18**:1-7

[10] Monroy Torres R, Ramírez Gómez X, Naves Sánchez J, Macías Hernández AE. Accesibilidad a agua potable para el consumo y preparación de alimentos en una comunidad expuesta a agua contaminada con arsénico. Revista Médica de la Universidad Veracruzana. 2009;**9**(suppl 1):10-13. Available from: http://www.uv.mx/rm/num_anteriores/revmedica_vol9_num1/suplemento/suplemento.pdf [Accessed: January 13, 2018]

[11] Kapaj S, Peterson H, Liber K, Bhattacharya P. Human health effects from chronic arsenic poisoning: A review. Journal of Environmental Science and Health Part A Toxic/Hazardous Substances and Environmental Engineering. 2006;**41**(10):2399-2428. DOI: 10.1080/10934520600873571

[12] Norma oficial mexicana. NOM-127-SSA1-1994, Salud ambiental, agua para uso y consumo humano-límites permisibles de calidad y tratamientos a que debe someterse el agua para su potabilización. Available from: http://www.salud.gob.mx/unidades/cdi/nom/127ssa14.html [Accessed: January 11, 2018]

[13] Nordberg GF, Fowler BA, Nordberg M. Handbook on the toxicology of metals. Estados Unidos: Academic Press; 2014

[14] Spayd SE, Robson MG, Xie R, Buckley BT. Importance of arsenic speciation in populations exposed to arsenic in drinking water. Human and Ecological Risk Assessment. 2012;**18**(6):1271-1291. DOI: 10.1080/10807039.2012.722824

[15] De Fátima Pinheiro Pereira S, Saraiva AF, de Alencar MI, Ronan SE, de Alencar WS, Oliveira GR, Silva CS E, Miranda RG. Arsenic in the hair of the individuals in Santana-AP-Brazil: Significance of residence location. The Bulletin of Environmental Contamination and Toxicology. 2010;**84**(4):368-372. DOI: 10.1007/s00128-010-9969-0

[16] Buchet JP, Lauwerys R, Roels H. Urinary excretion of inorganic arsenic and its metabolites after repeated ingestion of sodium metaarsenite by volunteers. International Archives of Occupational and Environmental Health 1981;**48**:111-118. Doi.org/10.1007/BF00378431

[17] Liu J, Waalkes MP. Liver is a target of arsenic carcinogenesis. Toxicological Sciences. 2008;**105**(1):24-32. DOI: 10.1093/toxsci/kfn120

[18] WHO. Arsenic and Arsenic Compounds. 2nd ed. Environmental Health Criteria 224. Ginebra: Environmental Health Criteria 224; 2001. Available from: http://www.who.int/ipcs/publications/ehc/ehc_224/en/ [Accessed: December 10, 2017]

[19] Caldwell KL1, Jones RL, Verdon CP, Jarrett JM, Caudill SP, Osterloh JD. Levels of urinary total and speciated arsenic in the US population: National Health and Nutrition Examination Survey 2003-2004. The Journal of Exposure Science and Environmental Epidemiology 2009;**19**(1):59-68. Doi: 10.1038/jes.2008.32

[20] Fillol C, Dor F, Labat L, Boltz P, Le Bouard J, Mantey K, Mannschott C, Puskarczyk E, Viller F, Momas I, Seta N. Urinary arsenic concentrations and speciation in residents living in an area with naturally contaminated soils. Science of the Total Environment. 2010;**408**(5):1190-1194. DOI: 10.1016/j.scitotenv.2009.11.046

[21] Maharjan M, Shrestha RR, Ahmad SA, Watanabe C, Ohtsuka R. Prevalence of arsenicosis in terai, Nepal. Journal of Health, Population, and Nutrition. 2006;**24**(2):246-252. Available from: http://www.jstor.org/stable/23499363 [Accessed: December 11, 2017]

[22] Nandana D, Somnath P, Debmita C, Nilanjana B, Niladri S, Nilendu S, Tanmoy JS, Santanu B, Saptarshi B, Papiya M, Apurba KB, States JC, Giri AK. Arsenic exposure through drinking water increases the risk of liver and cardiovascular diseases in the population of West Bengal, India. BMC Public Health. 2012;**12**(639):1-9. DOI: 10.1186/1471-2458-12-639

[23] Nandi D, Patra RC, Swarup D. Effect of cysteine, methionine, ascorbic acid and thiamine on arsenic-induced oxidative stress and biochemical alterations in rats. Toxicology. 2005;**211**(1-2):26-35. DOI: 10.1016/j.tox.2005.02.013

[24] Parvez F, Chen Y, Argos M, Hussain AZ, Momotaj H, Dhar R, van Geen A, Graziano JH, Ahsan H. Prevalence of arsenic exposure from drinking water and awareness of its health risks in a Bangladeshi population: Results from a large population-based study. Environmental Health Perspectives. 2006;**114**(3):355-359. DOI: 10.1289/ehp.7903

[25] ATSDR (Agency for Toxic Substances and Disease Registry). Toxicological Profile for Arsenic. Agency for Toxic Substances and Disease Registry, U.S. Public Health Service, Atlanta, GA. 1989

[26] Hultberg B, Andersson A, Isaksson A. Interaction of metals and thiols in cell damage and glutathione distribution: Potentiation of mercury toxicity by dithiothreitol. Toxicology. 2001;**156**(2-3):93-100. DOI: 10.1016/S0300-483X(00)00331-0

[27] Basu A, Mitra S, Chung J, et al. Creatinine, diet, micronutrients, and arsenic methylation in West Bengal, India. Environmental Health Perspectives. 2011;**119**(9):1308-1313. DOI: 10.1289/ehp.1003393

[28] Mitra SR, Guha-Mazumder DN, Basu A, Block G, Haque R, Samanta S, Ghosh N, Hira Smith MM, von Ehrenstein OS, Smith AH. Nutritional factors and susceptibility to arsenic-caused skin lesions in west Bengal, India. Environmental Health Perspectives. 2004;**112**(10):1104-1109

[29] Heck JE, Gamble MV, Chen Y, Graziano JH, Slavkovich V, Parvez F, Baron JA, Howe GR, Ahsan H. Consumption of folate-related nutrients and metabolism of arsenic in Bangladesh. The American Journal of Clinical Nutrition. 2007;**85**(5):1367-1374

[30] Gamble MV, Liu X, Ahsan H, Pilsner JR, Ilievski V, Slavkovich V, Parvez F, Chen Y, Levy D, Factor-Litvak P, Graziano JH. Folate and arsenic metabolism: A double-blind, placebo-controlled folic acid-supplementation trial in Bangladesh. The American Journal of Clinical Nutrition. 2006;**84**(5):1093-1101

[31] Gamble MV, Liu X, Slavkovich V, Pilsner JR, Ilievski V, Factor-Litvak P, Levy D, Alam S, Islam M, Parvez F, Ahsan H, Graziano JH. Folic acid supplementation lowers blood arsenic. The American Journal of Clinical Nutrition. 2007;**86**(4):1202-1209

[32] Zablotska LB, Chen Y, Graziano JH, et al. Protective effects of B vitamins and antioxidants on the risk of arsenic-related skin lesions in Bangladesh. Environmental Health Perspectives. 2008;**116**(8):1056-1062

[33] Maharjan M, Watanabe C, Ahmad SA, Umezaki M, Ohtsuka R. Mutual interaction between nutritional status and chronic arsenic toxicity due to groundwater contamination in an area of Terai, lowland Nepal. The Journal of Epidemiology and Community Health. 2007;**61**(5):389-394. DOI: 10.1136/jech.2005.045062

[34] Steinmaus C, Carrigan K, Kalman D, Atallah R, Yuan Y, Smith AH. Dietary intake and arsenic methylation in a U.S. population. Environmental Health Perspectives. 2005;**113**(9): 1153-1159. DOI: 10.1289/ehp.7907

[35] Monroy-Torres R, Macías-Hernández AE, Gallaga-Solórzano JC, Santiago-García EJ. Arsenic in Mexican children exposed to contaminated well water. Ecology of Food and Nutrition. 2009a;**48**(1):59-75. DOI: 10.1080/03670240802575519

[36] Monroy-Torres R, Velásquez A, Ortiz A. Programa Oportunidades sobre la Seguridad Alimentaria y Nutricional en Atarjea: Desde la percepción de sus participantes. Avances en seguridad alimentaria y nutricional. 2010;**1**(1):63-73. Available from: http://www.kerwa.ucr.ac.cr/bitstream/handle/10669/13417/1616-2436-2-PB.pdf?sequence=1&isAllowed=y [Accessed: December 02, 2017]

[37] Monroy-Torres R, Naves-Sánchez J, Melgar-Quiñonez H. Food security and Access to water in Mexican households. FASEB Journal. 2014;**28**(Suppl1):805-801

[38] Monroy-Torres R, Espinoza-Pérez A, Pérez-González RM. Evaluation of feeding and nutrition practices in a population exposed to arsenic: A proposal to integrate indicators of nutritional exposure. Nutrición Clínica y Dietética Hospitalaria. 2016;**2**(36):140-149. DOI: 10.12873/362monroytorres

[39] Monroy-Torres R, Espinoza-Pérez A. Factors that intensify the toxicological risk in communities exposed to arsenic in water. CienciaUAT. 2018;**12**(2):148-157. DOI: 10.29059/cienciauat.v12i2.803

[40] Monroy-Torres R et al. Effect of 4-week multivitamin supplementation on nutritional status and urinary excretion of arsenic in adolescents. Accepted in Journal of Nutrición Hospitalaria

[41] Vahter ME. Interactions between arsenic-induced toxicity and nutrition in early life. Journal of Nutrition. 2007;**137**:2798-2804. DOI: 10.1093/jn/137.12.2798

[42] Escudero Álvarez E, González Sánchez P. La fibra dietética. Nutrición Hospitalaria. 2006;**21**(Supl. 2):61-72. Available from: http://scielo.isciii.es/pdf/nh/v21s2/original6.pdf [Accessed: December 08, 2017]

[43] Massey VL, Stocke KS, Schmidt RH, Tan M, Ajami N, Neal RE, et al. Oligofructose protects against arsenic-induced liver injury in a model of environment/obesity interaction. Toxicology and Applied Pharmacology. 2015;**284**(3):304-314. DOI: 10.1016/j.taap.2015.02.022

[44] Aguilera Y, Liébana R, Herrera T, Rebollo-Hernandez M, Sanchez Puelles C, Benítez V, et al. Effect of Illimination on the content of melatonin, phenolic compouds, and antioxidant activity during germination of lentils (*Lens culinaris* L.) and kidney beans (*Phaseolus vulgaris* L.). The Journal of Agricultural and Food Chemistry. 2014;**62**(44):10736-10743. DOI: 10.1021/jf503613w

[45] Rivera JA, Muñoz-Hernández O, Rosas-Peralta M, Aguilar-Salinas CA, Popkin BM, Willet WC. Consumo de bebidas para una vida saludable: Recomendaciones para la población mexicana. Salud Publica Mex. 2008;**50**:171-193. Available from: http://www.medigraphic.com/pdfs/revinvcli/nn-2008/nn082i.pdf [Accessed: December 08, 2017]

[46] Gutiérrez JP, Rivera-Dommarco K, Shamah-Levu T, Villalpando-Hernández S, Franco A, Cuevas-Nasu L, Romero-Martínez M, Hernández-Ávila M. Encuesta nacional de salud y nutrición 2012. Resultados nacionales. Cuernavaca, México: Instituto Nacional de Salud Pública; 2012

Chapter 5

Mechanisms of Arsenic-Induced Toxicity with Special Emphasis on Arsenic-Binding Proteins

Afaq Hussain,
Vineeth Andisseryparambil Raveendran,
Soumya Kundu, Tapendu Samanta,
Raja Shunmugam, Debnath Pal and
Jayasri Das Sarma

Additional information is available at the end of the chapter

http://dx.doi.org/10.5772/intechopen.74758

Abstract

The importance of different arsenic forms in public health is well recognized owing to its distinct physical characteristics and toxicity. Chronic arsenic exposure has left a trail of disastrous health consequences around the world. However, the mechanisms behind the toxicity and the consequential diseases occurring after acute or chronic exposure to arsenic are not well understood. The toxicity of trivalent arsenic primarily occurs due to its interaction with cysteine residues in proteins. Arsenic binding to protein may alter its conformation and interaction with other functional proteins leading to tissue damage. Therefore, there has been much emphasis on studies of arsenic-bound proteins, for the purpose of understanding the origins of toxicity and to explore therapeutics. This book chapter illustrates the molecular mechanisms of arsenic toxicity with a special emphasis on arsenic binding to proteins and its consequences in alteration of tissue homeostasis.

Keywords: arsenic, gap junction intercellular communication (GJIC), gap junction proteins, connexin 43, DJ-1, sulfhydryl groups

1. Introduction

Long-term exposure to arsenic has resulted in the largest mass poisoning of the human population, making more than 100 million people defenseless against cancer and other arsenic-related diseases [1, 2]. Epidemiological studies have revealed that arsenic exposure

© 2018 The Author(s). Licensee IntechOpen. This chapter is distributed under the terms of the Creative Commons Attribution License (http://creativecommons.org/licenses/by/3.0), which permits unrestricted use, distribution, and reproduction in any medium, provided the original work is properly cited. (cc) BY

spans a wide geographical area spread across continents, with contaminations originating from soil, water, air and even food. Arsenic pollution gets aggravated through natural processes like volcanic eruptions, weathering, and biological activity. Anthropogenic activities, such as ore smelting, mining, well drilling and combustion of fossil fuels, also accelerate infusion of arsenic into places of human habitation [3]. Owing to its toxic nature, arsenic is a threat not only to humans but also to other living species. **Figure 1** illustrates natural and anthropogenic sources of arsenic.

Many mechanisms of arsenic-induced carcinogenicity have been proposed like DNA repair inhibition, oxidative stress, epigenetic events, effect on signal transduction and genotoxic damage. Studies have been focused to understand the molecular mechanisms of arsenic-induced carcinogenesis with an emphasis on oxidative stress and related signal transduction pathways. One of the hallmarks of oxidative stress is generation of reactive oxygen species (ROS) which triggers the antioxidant pathways as a cellular defense response. Two of the major players of cellular defence response on arsenic exposure are nuclear factor (erythroid-derived 2)-like 2 (Nrf2) and Parkinson's disease protein 7 (DJ-1), and their interplay results in activation and upregulation of several genes like glutathione-S-transferase A2 (GSTA2), NAD (P)H dehydrogenase quinone 1 (NQO1) and thioredoxin (Trx). There has been increasing evidence correlating arsenic exposure to reactive oxygen species (ROS) generation, DNA

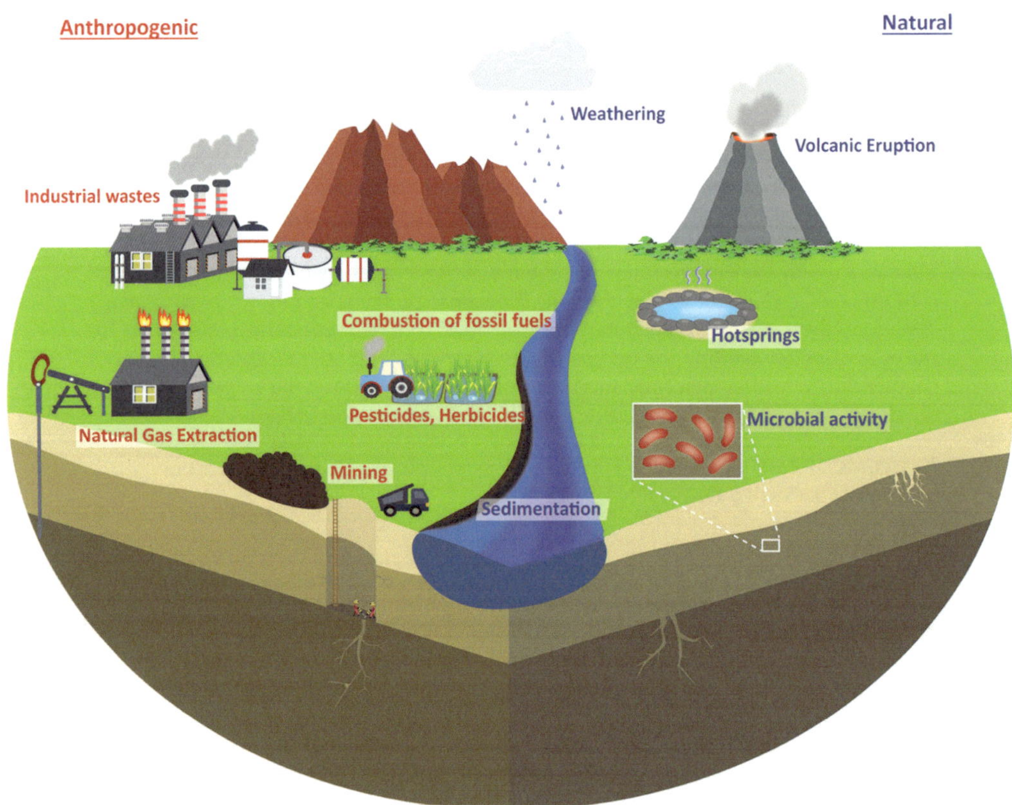

Figure 1. Mobilization of arsenic into environment.

damage and tumor promotion. Inorganic arsenic has been recognized as a potent human carcinogen. A number of epidemiological studies have found that human populations exposed to arsenic are prone to different types of cancers including that of the bladder, lung, skin, liver and kidney [4, 5]. Human body responds to arsenic ingestion through a set of concerted metabolic actions starting with methylation of the inorganic arsenic to monomethylarsonic (MMA^V) acid, which is then methylated again to dimethylarsinic acid (DMA^V) to permit its excretion through urine. However, this response may result in persistent methyl exhaustion in the event of chronic arsenic exposure leading to hypomethylation of DNA, which can alter the gene expression making the cells susceptible to carcinogenesis [6]. Interestingly, arsenic alone is considered to be a very weak mutagen; however, its synergistic association with genotoxic agents like ultraviolet radiation is reported to make it a potent mutagen [7]. Notwithstanding, the diverse mechanisms of arsenic toxicity need far greater elucidation, though the health hazards are well understood.

From the mechanistic standpoint, arsenic binding to cellular proteins can be a plausible mechanism of toxicity based on two hypotheses premised on functional disruption arising out of (a) sulfhydryl groups in proteins forming covalent bond with arsenite [8] and (b) the phosphate groups in proteins replaced by an arsenate. Arsenic binding to a specific protein could change the conformation and interaction with other functional proteins [9]. Therefore, many studies have been undertaken to examine the direct binding of arsenic to proteins, for the understanding mechanisms of arsenic toxicity and designing therapeutics against it.

All proteins with functionally important and conserved cysteine (Cys) residues, whose sulfhydryl groups are reactive nucleophiles or form disulfide bonds, are potential targets of functional disruption during chronic arsenic exposure. One such protein with conserved cysteines is the gap junction protein, connexin 43 (Cx43), belonging to the connexin family, and is the most commonly expressed member in different cell types. Our recent study showed that direct arsenic binding to this protein causes alteration in trafficking and the absence of gap junctional plaques on cell surface, resulting in propensity for cell proliferation. Given the hazardous nature of arsenic, the qualitative and quantitative analysis of arsenic is a much needed requirement. The conventional methods like neutron activation analysis and X-ray analysis, atomic absorption spectrometry (HG-AAS) and stripping voltammetry are very costly as well as complex. So, the quest for easy and cost-effective method continues till date. One such method gaining reputation in the field relies on optical sensors which have been discussed in this chapter. This chapter summarizes numerous traits of arsenic toxicity and emphasizes the interaction of arsenic with proteins to evaluate the chemical, biological, and physiological consequences.

2. Biogeochemical cycle: transformation and mobilization of arsenic in nature

Arsenic is commonly mobilized into the environment due to both natural and anthropogenic processes. The natural processes include geological (weathering of rocks and volcanic eruptions) and biological (microbial activity) events (**Figure 1**). Ancient or recent volcanic activities results in the inclusion of arsenic in the environment [10]. The earth's atmosphere also has a significant presence of arsenic species owing to wind erosion processes, sea spray, hot springs, volcanic emissions, forest fires and volatilization (in cold climates). Human activities like pharmaceutical manufacturing, glassmaking industry, wood processing, chemical weapons, burning of arsenic-rich fossil fuels and electronics industry also contribute to the addition of arsenic compounds into the environment [11]. Industrial by-products and wastes, ore smelting, mineral mining and well drilling can also mobilize and intensify arsenic into the environment.

Microbial metabolisms like arsenate reduction, arsenite oxidation and methylation processes are also a determining factor of the occurrence of the various arsenic oxidation states in the environment. Reduction of arsenate to arsenite by arsenate reductase enzymes is a common feature in the microbial world, while incidences of oxidation of arsenite to arsenate have also been reported in contaminated environments. These reactions also contribute to the protective and/or energy metabolisms of the bacteria from various arsenic-induced stress conditions (**Figure 2**) [12, 13].

3. Cellular mechanisms of arsenic toxicity

The levels of ROS play a key role in normal cell signaling, and its alteration can result in aberrant expression of genes that are activated by redox mechanisms. Notably, genes associated with redox mechanisms include those regulating cellular proliferation, differentiation and apoptosis. The consequences of ROS production can further lead to DNA damage which typically involves the conversion of 2-deoxyguanine to 8-hydroxyl-2′-deoxyguanosine (8-OHdG),

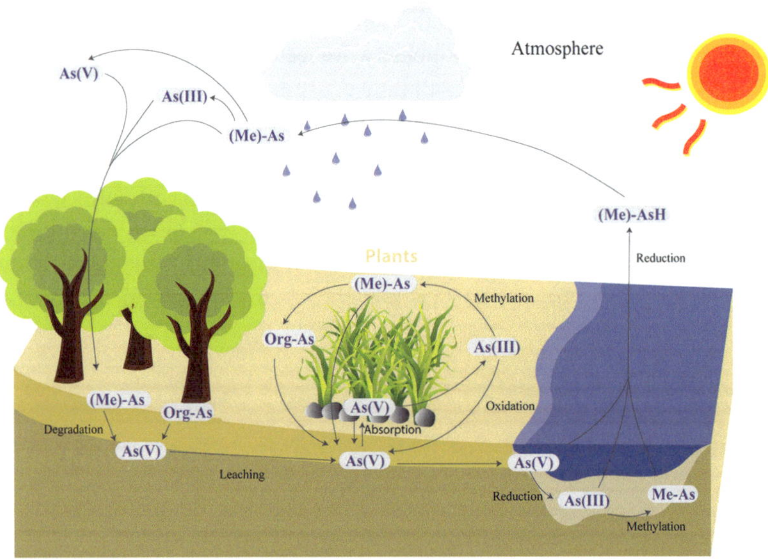

Figure 2. Biogeochemical cycle of arsenic.

which is considered as a marker indicating oxidative stress of DNA. Arsenic was capable of inducing specific DNA lesions consistent with oxidative damage like 8-OHdG generation. Moreover, 8-OHdG has also been detected in the skin of patients with arsenic-related Bowen's disease and in the liver of rats exposed to dimethylarsinic acid (DMA^{V}). These results indicate that ROS generation is a major pathway for arsenic-mediated genotoxicity in mammalian cells [14, 15].

Glutathione and other aminothiols such as cysteine and cysteamine comprise the non-protein sulfhydryls (NPSHs) in a cell and have significant free radical scavenging abilities. Therefore, depletion of intracellular glutathione levels is known to have an effect on arsenic mutagenesis. Studies have shown that pretreatment of cells with an inhibitor of glutathione biosynthesis (buthionine sulfoximine) reduces NPSH levels in the cell, resulting in enhancement of both the cytotoxicity and mutagenicity of arsenic. In contrast, glutathione and cysteine pretreatments are capable of protecting mammalian cells against the toxic effects of arsenite [16].

In a similar way, various antioxidants also have a significant effect on arsenic-induced genotoxicity. The balance between the rate of generation of free radicals and the rate of their removal by various antioxidant enzymes dictates the deleterious effect of oxidative stress. Enzymes like superoxide dismutase (SOD) and catalase are capable of partially suppressing both the toxicity and the mutagenic potential of sodium arsenite. These enzymes catalyze the dismutation of superoxide anions and prevent the formation of hydroxyl radicals by removal of hydrogen peroxide, respectively. Therefore, catalase and SOD are capable of reducing the mutagenic potential of arsenic. This is also consistent with other reports which reveal the ability of sodium arsenite to induce heme oxygenase, an oxidative stress protein, and peroxidase in various human cell lines. Moreover, the arsenite-induced occurrence of sister chromatid exchanges is reduced by SOD in cultured human lymphocytes [16].

In mammalian liver, the methylation of arsenic to MMA and DMA occurs at a high level by an incompletely characterized methyltransferase (**Figure 3**) using S-adenosylmethionine (SAM) as a methyl donor. SAM is a global methyl donor, required for DNA methylations, and its depletion can lead to hypomethylation of DNA resulting in alteration of gene expression like c-Myc, c-Met, cyclin D1 and induction of carcinogenesis [17, 18].

DNA methylation is an epigenetic modification that plays an important role in controlling the expression of various genes. Methylation generally occurs at cytosine residues located in symmetrical CpG nucleotide sequences, and its alteration, both in the global and regional levels, has been associated with oncogenesis. Methylation of CpG islands in the promoter region suppresses gene expression, as 5-methylcytosine interferes with the binding of transcription factors or other DNA-binding proteins causing reduced transcription. On the other hand, promoter hypomethylation causes overexpression of associated genes. Therefore, aberrant DNA methylation could be an underlying epigenetic mechanism causing altered gene expression that contributes towards the formation of cancers. This has been studied well in hepatocytes where chronic arsenic exposure induces hepatic DNA hypomethylation, which can potentially lead to aberrant gene expression and oncogenic growth in the liver, therefore suggesting a plausible mechanism of hepatocarcinogenesis (major cellular effects of arsenic are summarized in **Figure 4**) [18].

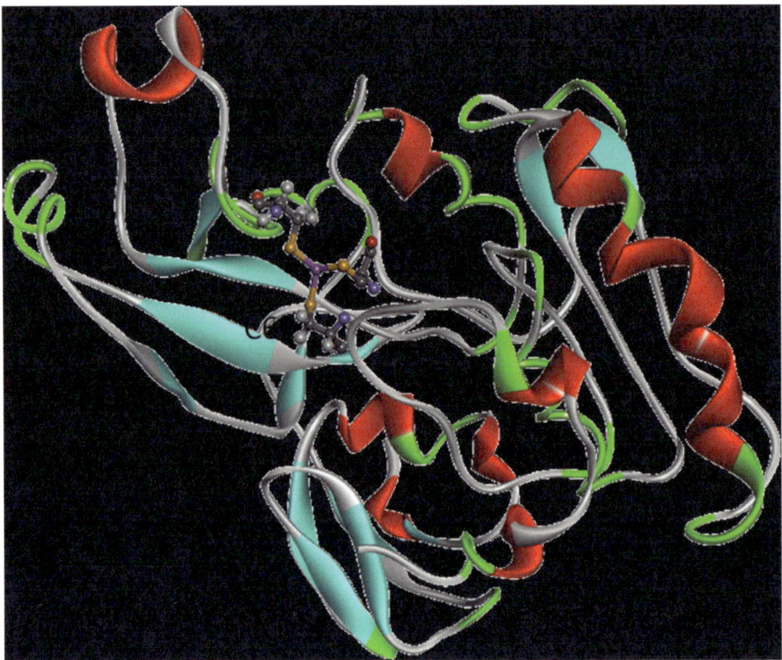

Figure 3. A homology model for arsenite methyltransferase from humans (AS3MT_HUMAN) showing arsenic bound to Cys residues. PDB ID: 5EVJ with 42% sequence identity spanning residues 38–327 was used to build the model. The coordinates were downloaded from https://swissmodel.expasy.org/repository/uniprot//Q9HBK9 and refined to introduce the arsenic atom.

Estrogens are considered to be liver carcinogens in rodents and are suspected to cause carcinogenesis in humans [19]. Evidence suggests that they cause hepatocellular proliferation and aberrant mitogenesis through ER-mediated mechanisms in addition to the likelihood that they confer epigenetic modifications. Hypomethylation of estrogen receptor-α (ER-α) promoter region caused by arsenic exposure and ER-α overexpression have been found to trigger associated formation of proliferative lesions and hepatocellular carcinogenesis. Therefore, chronic arsenic exposure causes overexpression of ER-α creating hypersensitivity of hepatic cells to endogenous steroids. As evidenced by microarray analysis, various cell cycle-regulating genes like cyclin D1, cyclin D2 and cyclin D3 were overexpressed on exposure to arsenic. Liver cells that acquired malignant properties upon arsenic treatment also showed cyclin D1 overexpression. In addition, this overexpression had a direct effect on the observed malignant transformation, as selective cyclin D1 overexpression in the liver was sufficient enough to initiate hepatocellular carcinogenesis. Cyclin D1 can, therefore, be considered as a hepatic oncogene. Cyclin D1 is also known to be upregulated transcriptionally by various growth factors which potentially include estrogens. In estrogen-responsive tissues like the liver and uterus, proliferative lesions and co-overexpression of ER-α and cyclin D1 after chronic arsenic exposure are reported. Cyclin D1 activation by arsenic may be a secondary effect to ER-α overexpression as cyclin D1 is potentially an ER-α-linked gene. Therefore, we can expect that aberrant expression of cyclin D1 along with that of other oncogenes leads to carcinogenic

http://dx.doi.org/10.5772/intechopen.74758

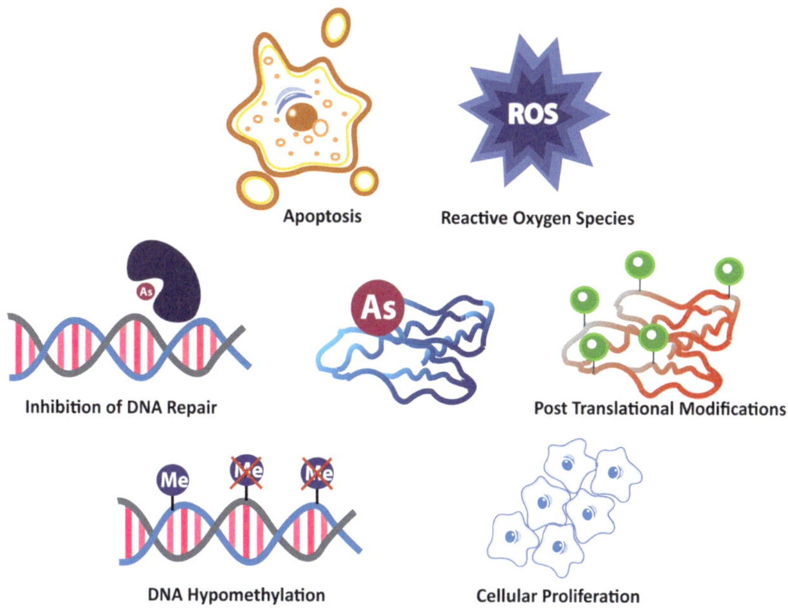

Figure 4. Cellular effects of arsenic toxicity.

transformation. Altogether, cyclin D1 overexpression was seen upon arsenic exposure in multiple in vitro and in vivo model systems of arsenic carcinogenesis, which includes skin and bladder cancers in rodents. Thus, under conditions of arsenic-induced carcinogenesis, overexpression of cyclin D1 is observed consistently [18–20].

In mouse lung tissue, reduced expression of proteins associated with cellular migration was observed when exposed to low dose of arsenic. On lung tissue of mice fed low-dose arsenic, changes in extracellular matrix (ECM) protein expression and a large increase in matrix metalloproteinase (MMP)-9 expression were revealed [21]. MMPs are responsible for ECM degradation among other proteolyses. MMP-9 is the most prominently studied MMP in the lung and has been associated with a variety of lung diseases [22]. An increase in the ratio of MMP-9 to tissue inhibitor of matrix metalloproteinase (TIMP)-1 was observed under low-level arsenic exposure [23]. This imbalance between MMP-9 and TIMP-1 can cause changes in epithelial wound response, thereby contributing to the progression of airway remodeling. Altered wound response is partly due to increased secretion and activity, upon increasing concentration of arsenic. Therefore, arsenic ingestion may alter wound response and, specifically, MMP-9/TIMP-1 ratios in the lung. To conclude, arsenic is capable of causing or exacerbating lung diseases by directly affecting signaling pathways involved in cell migration and remodeling of the airway [24].

Studies have revealed that both c-Jun NH2-terminal kinases (JNKs) and extracellular signal-regulated protein kinases (Erks) are activated by arsenite, with their activation varying temporally and depending on the dosage. Various results also indicate that Erk activation but not JNK activation is required for arsenite-induced cell transformation. Expression of the dominant-negative mutant JNK1 blocked induction of apoptosis by arsenite or arsenate compared with vector-transfected JB6 cells, indicating the role of activation of JNKs in arsenic-induced apoptosis. Studies have found that both arsenite and arsenate can cause transactivation of activation protein-1 (AP-1). Since increased activation of AP-1 by arsenite could be inhibited by either treating cells with MAP kinase Erk kinase (MEK)1 inhibitor or overexpression of dominant-negative protein kinase C α (PKCα), this induction appears to occur through activation of mitogen-activated protein (MAP) kinases and PKC. Moreover, in AP-1-luciferase reporter transgenic mice, transactivation of AP-1 was caused by both arsenite and arsenate. Recent data also indicates that PKC, upstream from the MAP kinases, may be involved in mediating arsenite-induced signal transduction. Activation of PKC requires it to be translocated from the cytosol to the membrane, and this phenomenon is observed within 15 minutes when cells are treated with arsenite. Moreover arsenite-induced AP-1 activity, phosphorylation of Erks, JNKs and p38 kinase were blocked once PKC activation was inhibited. These results suggest that PKC plays a critical role in arsenite-induced activation of MAP kinases [25, 26].

Nuclear factor kappa B (NF-κB) is a rapidly induced stress-responsive transcription factor that may play an important role in arsenic-induced signal transduction, cell transformation and apoptosis [27]. Reports suggest that in arsenic-induced oxidative stress, H_2O_2 and superoxide are the predominant reactive species in endothelia cells and may be the mediators for the activation of the NF-κB pathway. It was also shown that arsenic could induce activation of NF-κB in different cell culture models. Expression of a dominant-negative inhibitory kappa-B-α blocked arsenic-induced activation of NF-κB and apoptosis [26].

4. Arsenic binding to proteins

The trivalent arsenite has a tendency to bind to sulfhydryl groups. The cysteine residues are a direct target of arsenite in proteins and peptides [28]. The chemical reaction involved in arsenic binding to cysteines has been well recognized. Some of the chemicals like arsine halides used in warfare during the First World War owe their toxicity to their ability of binding to protein dithiols. To defy the toxic effects of such warfare agents, the British government approved the use of β-chlorovinyldichloroarsine (dithioglycerol) which has the ability to form stable complexes with arsenic. The competitive binding of arsenic to dithioglycerol rescues cellular proteins from binding to arsenic [29, 30].

Arsenic affinity for proteins can result in conformational changes in the protein and loss of protein–protein and protein-DNA interactions. *Escherichia coli* consists of a repressor protein ArsR in which each subunit within its α-helix contains two cysteine residues. The unraveling of this α-helix is required in order to accommodate trivalent arsenite for binding to the protein. The unraveling of the helix causes the conformational change in the protein that dissociates ArsR from DNA resulting in induction of gene expression. Arsenite binds to three cysteine (Cys32, Cys34 and Cys37) residues in ArsR, where Cys32 and Cys37 are present in the α-helix of the DNA-binding sites (**Figure 5**). The Cys residues in the protein are located in such a way that arsenite is unable to bind unless the protein unwinds for a conformational change [31].

http://dx.doi.org/10.5772/intechopen.74758

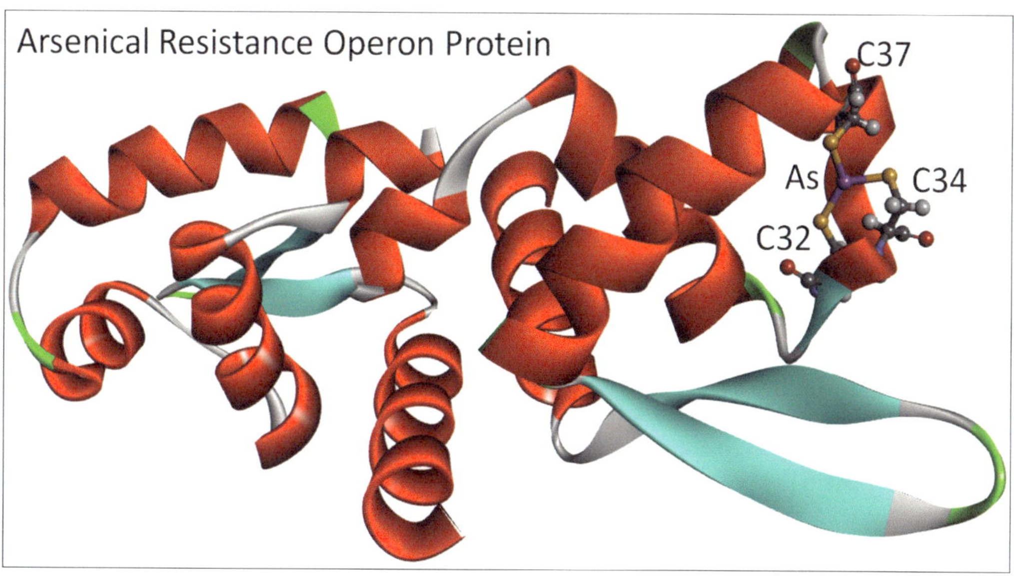

Figure 5. A homology model for arsenical resistance operon repressor protein from *E. coli* (ARSR_ECOLI) showing arsenic bound to Cys residues. PDB ID: 1SMT with 40% sequence identity spanning residues 8–90 was used to build the model. The coordinates were downloaded from https://swissmodel.expasy.org/repository/uniprot//P37309 and refined to introduce the arsenic atom.

4.1. Binding sites of arsenic in proteins

Cysteine and histidine residues are thought to be the most frequent targets of metals like zinc, copper and iron resulting in such metals binding to peptides and proteins [32, 33]. However, the binding of arsenic to histidine is not well understood and is yet to be established. There is no change in nuclear magnetic resonance (NMR) spectra once arsenic was added to a buffered solution of histidine signifying the absence of interaction between histidine and arsenic [34]. Many studies have also used site-directed mutagenesis to replace cysteine residues with serine residues on the reason that interaction between arsenic and serine is very weak. Arsenic is however known to bind to zinc finger protein in C3H1 motif and not in C2H2 motif, releases zinc, and thus decreases the capacity of the protein to bind to DNA. Selenocysteine—a cysteine analogue—also has the ability to bind to arsenic species. This amino acid is present in selenoproteins and has a lower pKa which increases nucleophilicity. The amino acid residues in the vicinity of cysteine (or selenocysteine) act as proton donors [35–37].

Studies with some enzymes reveal that serine residues can be potential targets of arsenic species, thereby inhibiting their function. It was found that in serine hydrolases and the arsenic moieties interacting with hydroxyl containing serine, pentavalent forms of arsenic rather than trivalent forms were prevalent. The complex between the serine residue and the pentavalent arsenic consists of a tripartite oxyanion hole in the proximity of the active site [38].

4.2. Arsenic binding to specific proteins

4.2.1. *Arsenic binding to hemoglobin*

Arsenic species are cleared from the blood immediately in humans, but the time of clearance of arsenic from animal species varies noticeably. The retention of arsenic in rat blood is longer when compared with other species. Arsenic has been found to bind to transferrin in hemodialysis patients [39]. Hemoglobin in red blood cells (RBCs) were predicted to be the sites of arsenic accumulation, because hemoglobin constitutes 97% of dry weight of RBCs [40]. The affinity of hemoglobin in rat liver is much higher in rats as compared with humans. The rat and human hemoglobins are tetramers, each consisting of two α-chains and two β-chains. The difference lies in the number of the cysteine residues with rat hemoglobin consisting of three cysteines (Cys111, Cys104 and Cys13) in α-chain, while two cysteines (Cys125 and Cys93) in β-chain. On the other hand, human hemoglobin has only one cysteine in α-chain and two cysteines in β-chain [3].

4.2.2. *Arsenic binding to glutathione*

The metabolism of arsenic in the cells involves the reduction of pentavalent arsenic to trivalent arsenic. This reaction consists of a redox cycle involving a bio-thiol (glutathione) with the production of a *tris*-glutathionyl-arsenite species. The multiple methylations of arsenite by *S*-adenosylmethionine to the generation of trimethylarsine (hemolytic toxin) also involve glutathione. Glutathione presence in the intermediate conjugate forms of methylated arsenic species helps these molecules to be removed from the cells by the multidrug-resistant proteins (having ATP-binding cassette). Dimethylarsinic acid (carcinogenic end-metabolite) also reacts with glutathione having a high cytolethal effect on cells. Moreover, various enzymes and regulatory elements can contribute to the arsenic biotransformation by contributing individual or multiple cysteine thiol groups in vicinity in proteins, for example, thiol groups required for catalytic activity [41].

4.2.3. *Arsenic binding to metallothioneins*

Metallothioneins are expressed by various organisms including bacteria, fungi, plants and vertebrates. They belong to a protein family of ubiquitous nature characterized by low molecular weight, high metal and cysteine content. They are capable of binding essential metal ions (zinc, copper) and toxic heavy metals (arsenic, cadmium).

Studies have revealed that bioaccumulation of arsenic in seaweed species *Fucus vesiculosus* is achieved through the binding of arsenite to the cysteine-rich metallothioneins. Moreover, arsenic is also known to bind to mammalian metallothioneins in rabbit and human species. It is present abundantly in the kidneys and liver of mammals. Further studies on human metallothioneins were consistent with the hypothesis that arsenite has a binding preference for three vicinal thiol groups, with α and β domain of human metallothionein containing 11 and 9 cysteines, respectively. All the 9 cysteines were involved in binding to three arsenite molecules in β domain, while in the case of α domain, only 9 out of 11 cysteine residues were involved in binding to three arsenites. This leaves two cysteine residues protonated with no fourth arsenite engaged in binding [42, 43].

4.2.4. Arsenic binding to ArsD As(III) metallochaperone

Arsenic being the most common toxic element in the environment has resulted in the evolution of arsenic detoxifying mechanisms in nearly all organisms. In archaea and bacteria, trivalent metalloids like arsenite are pumped out of the cell by ArsAB ATPases encoded by various *ars* operons. Three conserved cysteine residues (Cys12, Cys13 and Cys18) are required for the chaperone activity of ArsD. ArsD also helps to increase the arsenite affinity of Ars A enabling the detoxification of arsenite, even at low concentrations. In the case of ArsA, there are two cysteines (Cys113 and Cys422) in the high affinity metalloid-binding site along with the third cysteine that participates in activation of ATP hydrolysis. In the absence of arsenite, a low basal rate of ATPase activity is shown by ArsA [44].

4.2.5. Arsenic binding to other proteins

Trivalent arsenic species are also known to bind to other proteins like actin, tubulin, estrogen receptor and glucocorticoid receptors. Arsenite can bind to Kelch-like ECH-associated protein 1 (KEAP 1). This is a major antioxidant-sensing protein which acts at low K_d values. One of the most common motifs present in many proteins consists of two cysteine residues separated by two amino acids (CXXC). The presence of cysteine residues increases with increasing complexity of the organisms, making humans vulnerable to arsenic toxicity because of the high affinity of arsenic for cysteine residues.

One such important cell surface protein consisting of highly conserved cysteine residues is connexin 43 (Cx43)—a widely expressed gap junctional protein important for cell death, proliferation and differentiation [45]. Cx43 has nine Cys residues, six of which are in the extracellular domain and three in the intracellular domain. Six connexin monomers form a hemichannel called connexin. In the plasma membrane, one connexin can dock to another connexin in the plasma membrane of an adjacent cell resulting in the formation of complete gap junction channel. A hemichannel formed by single type of connexin isoforms is called homomeric hemichannel or consists of multiple types of isoforms called heteromeric hemichannel. Two identical homomeric or heteromeric hemichannels dock to form a homotypic channel, and two different homomeric or heteromeric hemichannels dock to form a heterotypic channel (**Figure 6**). Recent in silico studies (Hussain et al., manuscript communicated) in combination with cellular and biochemical analysis revealed insights into the binding modes of arsenite to conserved Cys groups in Cx43. In Cx43, As^{+3} can be bound to three cysteines in the intracellular domain in a monovalent fashion as they are free cys, while it can bind the extracellular domain cysteines in either monovalent, divalent or trivalent fashion depending on the state and location of the protein in the cell. Arsenite ion (As^{+3}) can attack the free sulfhydryl group until all the valencies of the As^{+3} are satisfied by covalent bonding to the sulfur from the cys residues. This profoundly affects the Cx43 primary, secondary, tertiary and the quaternary structure. This study is the first of its kind which shows that arsenic can directly bind to Cx43 via its highly conserved cysteine residues causing misfolding of Cx43, which leads to alteration of transportation, localization and oligomerization of Cx43. Further experiments revealed that Cx43 was colocalizing with ER marker (calnexin), revealing the inability of Cx43 to be transported beyond endoplasmic reticulum/endoplasmic reticulum

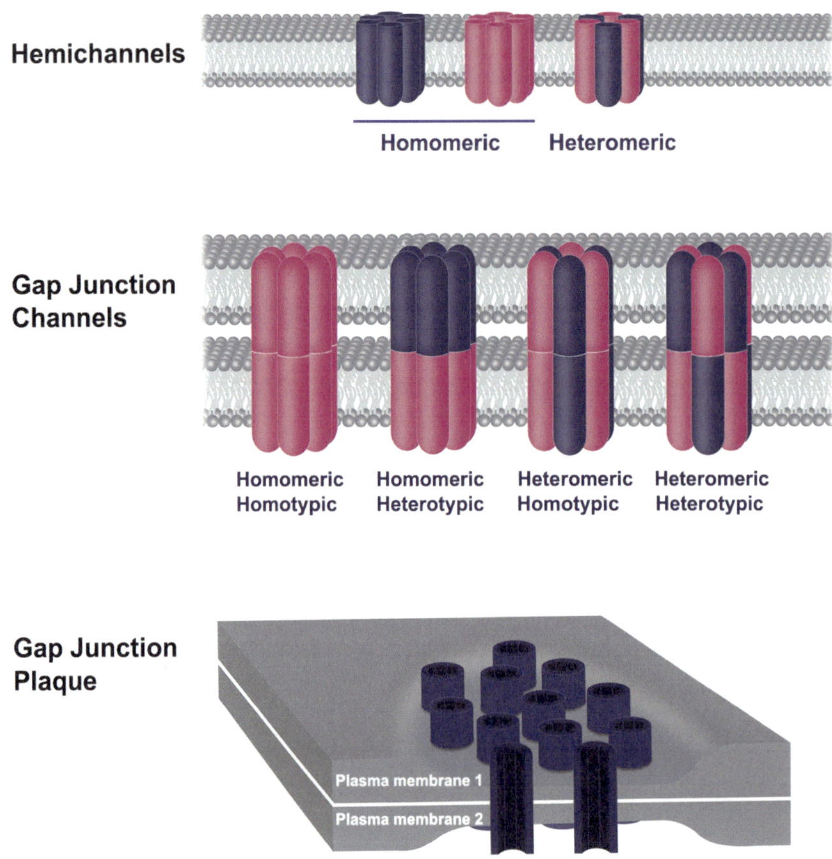

Figure 6. Hierarchy of structures involved in the formation of gap junction intercellular communication (GJIC).

Golgi intermediate compartment (ER/ERGIC) (Hussain et al., manuscript communicated). This loss of Cx43 composed of functional gap junctions on the cell surface has deleterious effect on cellular homeostasis (**Figure 7**).

Arsenic is considered a group 1 carcinogen by the International Agency for Research on Cancer (IARC) and causes cancers of the lung, liver and skin [46]. Gap junction intercellular communication has been found disrupted in many tumors and malignancies. Gap junctions are considered tumor suppressors, and the persistent downregulation of gap junction proteins makes cells susceptible to cancer [47]. Decreased or diminished expression and/or function of Cxs has been observed in most tumor cell lines and in solid tissue tumors, including melanomas. Our study revealed that arsenic causes disruption of gap junction intercellular communication both in vivo and in vitro. Arsenic is considered a weak mutagen; therefore, recent trends in the field have focused on deciphering the role of non-mutagenic pathways like cell–cell communication in arsenic-induced cancer. Our study revealed that arsenic induces disruption of gap junctions which are considered as tumor suppressors, thereby putting forward new non-mutagenic pathways which may be altered during the course of arsenic-induced carcinogenesis.

http://dx.doi.org/10.5772/intechopen.74758

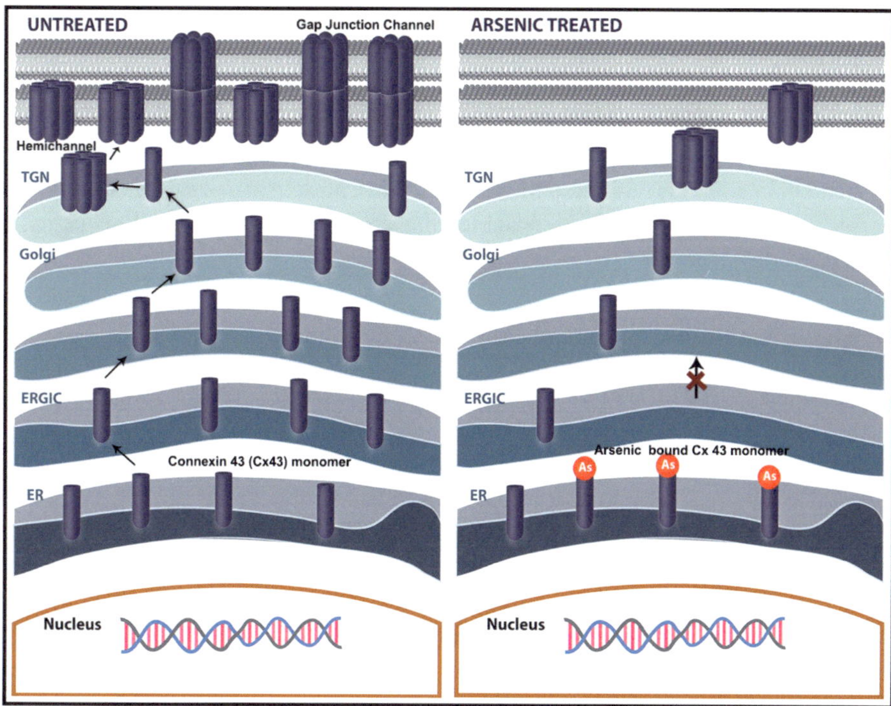

Figure 7. Arsenic binding causes alteration in trafficking of connexin 43 to the cell membrane.

Another such important cellular factor involved in cellular stress is DJ-1. DJ-1 is a 20KDa, homodimeric protein containing a nucleophilic elbow forming the active site of the protein. There are three important and conserved cysteine residues (Cys46, Cys53 and Cys106) in the DJ-1 protein, of which Cys53 and Cys106 are exposed. Cys106 has been found to be a prominent player in the nucleophilic groove that binds to divalent ions like zinc (II), copper (II) [48] and mercury [49] in vitro. Interaction with metal ions might be a possible mechanism of DJ-1-mediated cellular protection against metal-induced toxicity. Arsenic in the form of arsenite (As (III)) has been found to interact with three thiol group of cysteine residues [3]. Therefore, there is a possibility that arsenic binds to the nucleophilic groove in the homodimer of DJ-1. Oxidation state of the Cys106 is one of the determining factors behind the activity of the protein. Cysteine has the propensity to bind to three oxygen atoms resulting in the formation of the three forms—SOH, SO_2H and SO_3H. The presence of the SOH and SO_2H form activates the protein causing its translocation into the nucleus. Upon activation, DJ-1 regulates the activity of several transcription factors like nuclear factor erythroid 2-related factor 2 (Nrf2), polypyrimidine tract-binding protein-associated splicing factor (PSF) and sterol regulatory element-binding protein (SREBP), signal transducer and activator of transcription 1 (STAT1) and Ras-responsive element-binding protein (RREB1). DJ-1 has been found to inhibit phosphatase and tensin homolog (PTEN), an inhibitor of the AKT (protein kinase B) signaling pathway, resulting in enhanced cell proliferation. DJ-1 also functions in the sequestration of the death domain-associated protein (DAXX) in the nucleus. DAXX is required in the cytoplasm for providing the second activation signal to the phosphorylated apoptosis signal-regulating kinase 1 (ASK1) protein, which then triggers the apoptotic pathway. As a result, unavailability of DAXX in the cytoplasm hinders the initiation of the apoptotic pathway. Under condition of excess oxidative stress, the SO_3H form prevails which inactivates the protein and re-translocates back to the cytoplasm. As a result, the entire antioxidant response regulated by the activated DJ-1 protein is inhibited. Moreover, DAXX protein also becomes free, which then translocates into the cytoplasm and provides the required second activation signal to the phosphorylated ASK1 protein.

4.3. Therapeutic applications of arsenic binding to proteins

Arsenous acid [$As(OH)_3$] formed by dissolving of arsenic trioxide (As_2O_3) was found to be an effective and safe treatment for acute promyelocytic leukemia (APL) in the 1970s. The United States Food and Drug Administration approved the use of As_2O_3 as a treatment for APL in September 2000 [50]. As_2O_3 treatment was shown to have a dual effect on APL cells, with low arsenic concentration (0.25–0.50 mM) favoring APL cell differentiation and high concentrations (1–2 mM) inducing apoptosis (programmed cell death). Direct arsenic binding to cysteine residues present in zinc fingers of promyelocytic leukemia fusion protein (PML-RARa) was found to be a mechanism underlying APL remission [51]. Arsenic binding induces a conformational change in the structure of the protein (PML-RARa), facilitating its oligomerization. This oligomerization enhances ubiquitylation and SUMOylation, resulting in its degradation [52]. Given the constant requirement for DNA and protein synthesis, thioredoxin (Trx) and thioredoxin reductase (TrxR) are observed to be overexpressed in various tumors. Moreover, in vivo data suggests that TrxR is necessary for the growth of tumor cells, making them plausible targets for anticancer therapies [53].

Arsenic has been proposed to induce cell death through thioredoxin reductase (TrxR) inhibition, with both N-terminal dithiols and C-terminal selenothiol interacting with arsenic compound [54]. Sensitivity of cells to arsenic can be attributed to high expression of membrane transporter aquaglyceroporin which allows arsenite uptake, along with a low, basal level of cellular glutathione. Multiple factors such as liver damage, cardiac toxicity and peripheral neuropathies caused by toxicity at higher dosage of As_2O_3, along with bioavailability of arsenic compounds, limit the widespread use of As_2O_3 against solid tumors [50].

5. Arsenic sensing

Considering the hazardous facts of arsenic, it is very important to detect arsenic both qualitatively and quantitatively. Many conventional methods like hydride generation atomic absorption spectrometry (HG-AAS), neutron activation analysis and X-ray analysis and stripping voltammetry are available to determine arsenic. Though these methods are available, they are not very cost-effective and are very complex [55–59]. To determine arsenic, easy and cost-effective methods are yet to be explored. In recent times various heavy metals and toxic anions are detected selectively and sensitively by using optical detection techniques (fluorescence and UV–Vis), which implement a viable and simple approach towards

http://dx.doi.org/10.5772/intechopen.74758

the detection process. Except optical (fluorescence and UV–Vis) detection methods, other available methods need complex experimental setup; hence, they are far from 'on-field' application purpose. Simplicity, low-cost and 'on-field' application possibilities make optical sensing technique versatile.

Optical sensors can be of different types depending upon the material used for sensing. The first one is nanomaterial-based assays for the detection of the arsenic in different mediums. Though the detection of arsenic is tough, but researchers are able to draw an outline about the ligands which can bind arsenic, and these ligands can be used as a binding unit in a sensing material which leads to either color change or change in emission spectrum. As arsenic is very much labile towards thiol group, a bunch of thiolated ligands are reported for arsenic binding. These ligands are dithiothreitol (DTT), reduced glutathione (GSH) and cysteine, and **Figure 8** describes the chemical structure of these three ligands. Arsenic can bind with GSH and cysteine by forming As-O bond also, if no free –SH available. Except thiolated ligands, there are some ligands like humic acid [60] and N-(dithiocarboxy)-N-methyl-D-glucamine [61] which can also bind As(III) by forming As-O bond. Keeping this information in mind, gold nanoparticle-based sensors were reported for As(III) detection. The surface of the gold nanoparticles can be modified by the thiolated ligands, which after binding with As(III) showed a drastic color change to indicate the presence of the toxicant in the aqueous medium [62]. Aptamer-conjugated nanoparticles are also very effective composites which can detect arsenic in aqueous medium [63, 64] by changing the color. In all these types of detection assays, aggregation of the nanoparticles is the predominant factor to show the color change. Though these kinds of materials are responsive towards arsenic, but sensitivity is one of the issues which prevent these from field effectiveness.

Both selectivity and sensitivity are important for effective detection of arsenic. Small molecules are developed to detect different forms of arsenic in aqueous medium having good selectivity over other toxicants as well as good sensitivity. Baglan M et al. have reported a cysteine-fused tetraphenylethene, which can bind with As^{3+}, and showed aggregation-induced emission as a signal [65]. Here, also the thiol group of cysteine acts as the dominating factor for As^{3+} binding and leading to the close proximity arrangement of the tetraphenylethene. More toxic As^{3+} can be distinguished over less toxic As^{5+} using this system, and the detection limit tends to 0.5 ppb, which is lower than the limit according to the World Health Organization (WHO) [66]. Keeping besides the thiol systems, Somentah et al. have designed a simple Schiff base system which can identify the most toxic AsO_3^{3-} fluorometrically. 'Off–on' system in fluorescence is always most exciting and effective for the detection of pollutants. In this work they have designed a molecule which is initially not showing any fluorescence emission, but after selective addition of AsO_3^{3-} fluorescence, signal is turned on due to intermolecular H-bonding leading to chelation-enhanced fluorescence (CHEF) [67]. Development of arsenic sensor is evolving year

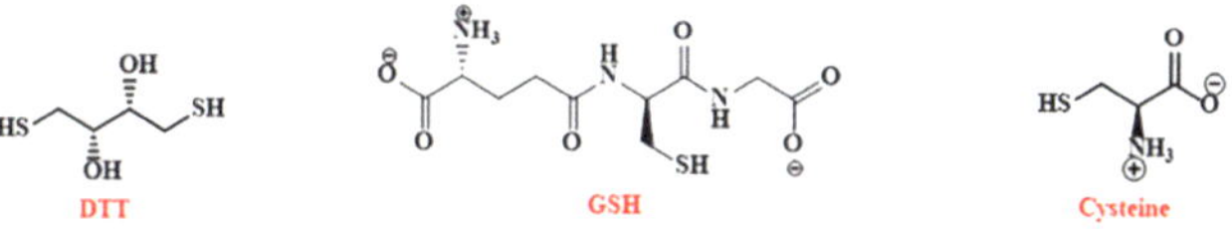

Figure 8. Chemical structures of thiolated ligands (DTT, GSH and cysteine).

after year due to the need of arsenic detection. A modified coumarin derivative was documented as an As^{3+} sensor having a detection limit of 0.53 nM. Though the system has excellent sensitivity, but the main drawback is its incapability of detecting As^{3+} in aqueous media. So, the sensing system which can work effectively in aqueous media for the detection of arsenic having fluorescence property is in tremendous search till date. In search of a suitable aqueous medium arsenic sensor, an inorganic co-crystal has been reported having a unique luminescent response to detect As(III), having a detection limit of 49 pM. But these types of systems are not that much useful for real-life application [68]. **Table 1** is prepared where available optical sensors are summarized.

A few small molecule sensors have been explored over the years, but the 'on-field' application is quite tough for small molecule sensors due to their low molecular weight and water solubility. To overcome such issues, polymeric sensing assays are developed as they have high molecular weight, tunable solubility by introducing hydrophilic functionality, high signal amplification and high sensitivity due to the number of more repeating units. In the field of materials science research, polymer-based substances have high priority. For sensing of arsenic, polymer-based sensing assay is very rare, with a few number of reports existing. A pyridylmethyl-appended 2-aminothiophenol with 2,6-diformyl-4-methylphenol was developed, which can detect arsenate (As(v)) selectively in aqueous medium. But the interesting fact

Types of sensors	Compounds or materials	Mode of sensing with optical changes
1.Nanoparticle-based	1. 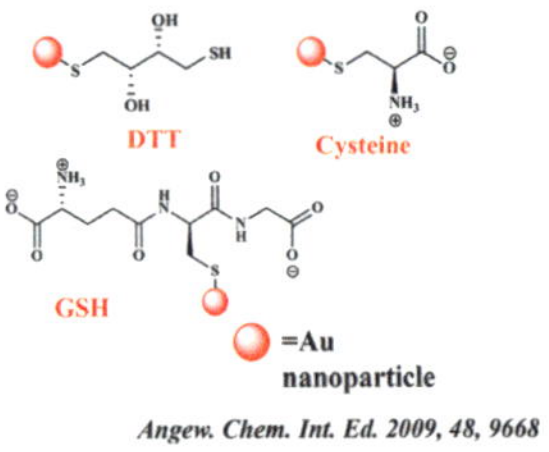 DTT; Cysteine; GSH; =Au nanoparticle; *Angew. Chem. Int. Ed. 2009, 48, 9668*	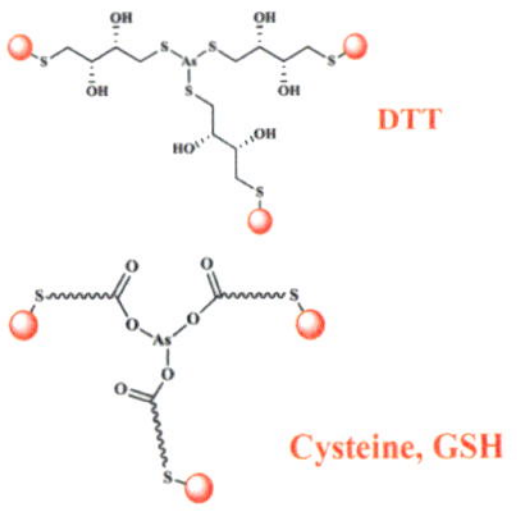 DTT; Cysteine, GSH
	2. 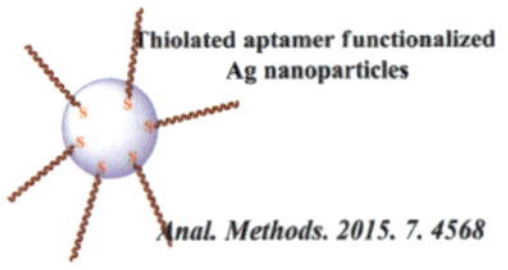 Thiolated aptamer functionalized Ag nanoparticles; *Anal. Methods. 2015. 7. 4568*	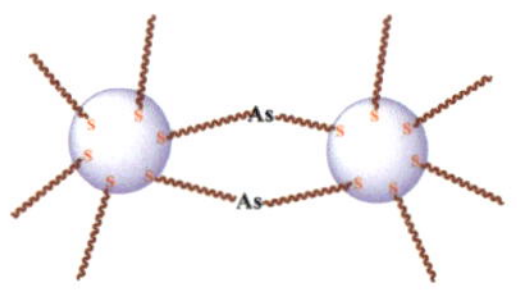
	3. 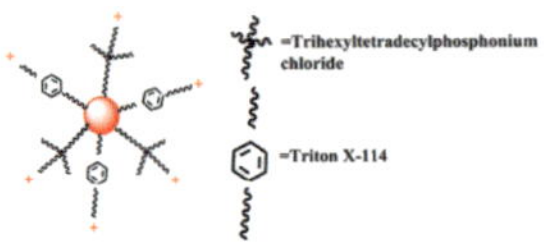 =Trihexyltetradecylphosphonium chloride; =Triton X-114	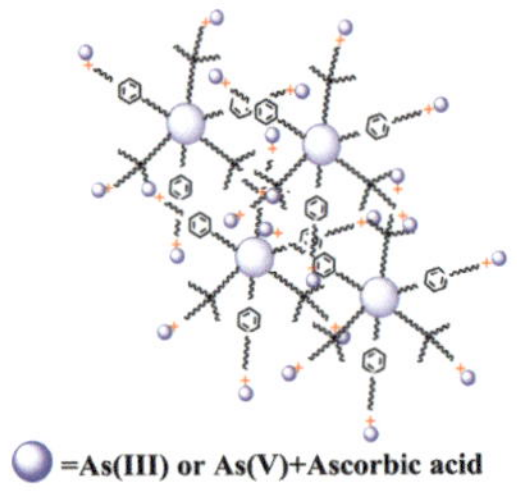 =As(III) or As(V)+Ascorbic acid

Types of sensors	Compounds or materials	Mode of sensing with optical changes
2. Small molecule-based	1. 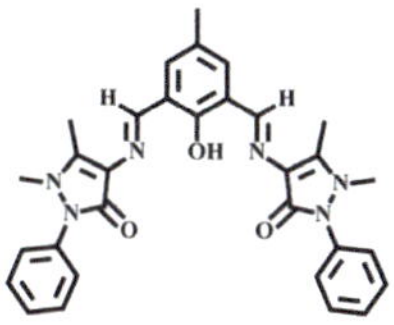	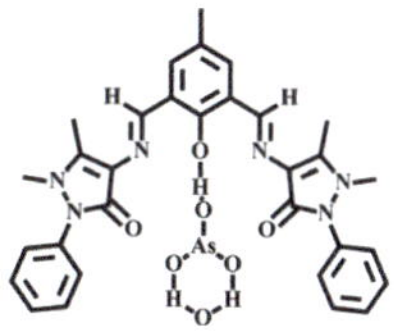
	2. 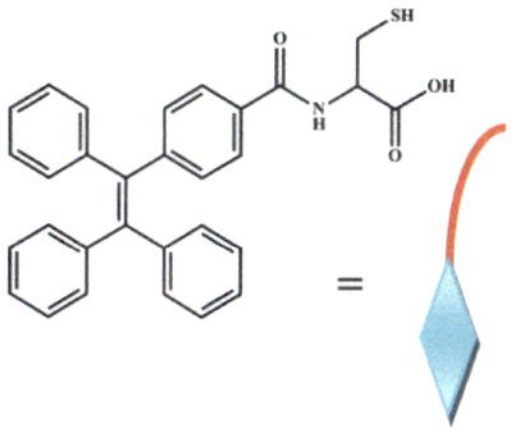	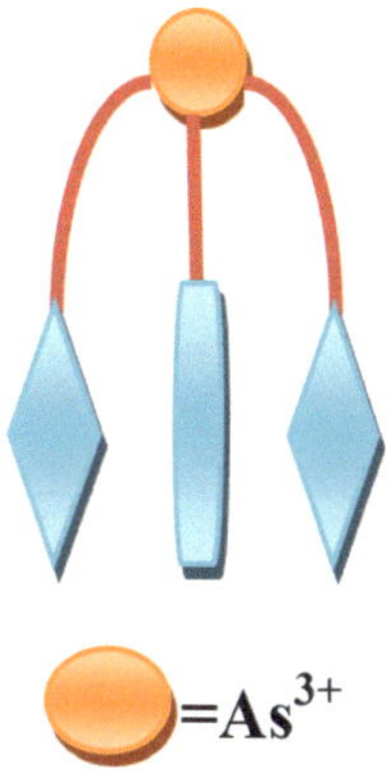 =As^{3+}
3.Polymer-based	1. 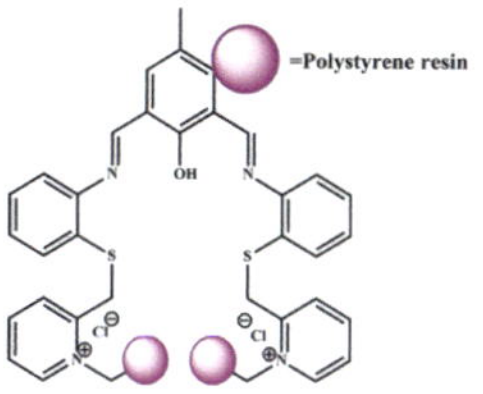 =Polystyrene resin	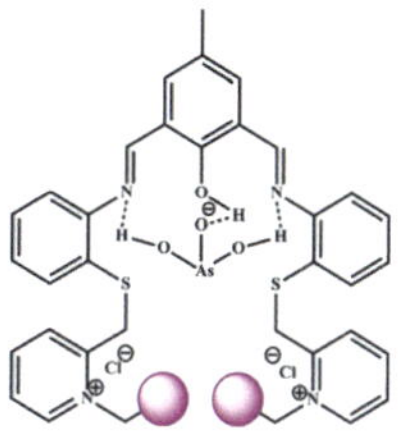
	2. 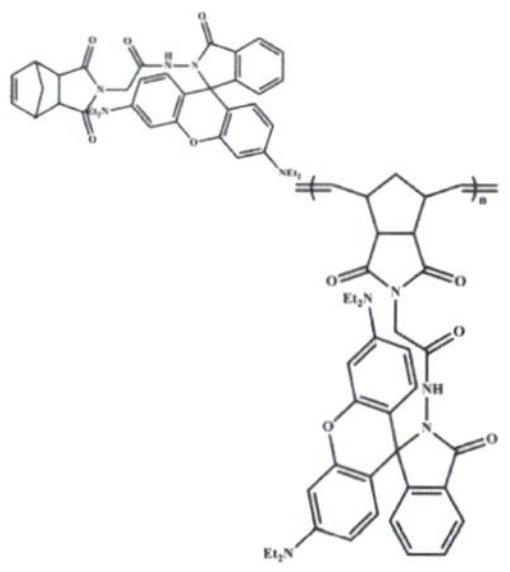 J. Mater. Chem. A, 2013, 1, 8398	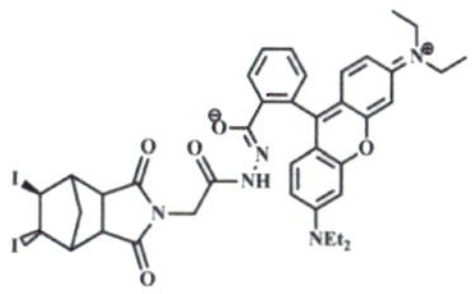

Table 1. Some available optical arsenic sensors and their mode of detections.

is that after attachment of the small molecule with polystyrene resin, the new material consists both sensing and removal property of As(V) which is very beneficial for the treatment of As(V) in drinking water practically [69]. All mentioned sensory assays are responding due to

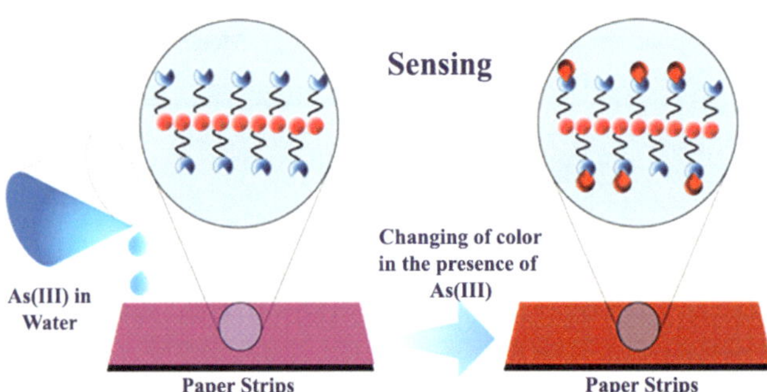

Figure 9. Cartoon representation to demonstrate change in color of PNor-Rh-coated paper strip in the absence and presence of As(III).

interaction of host and guest. But one of the best indirect As(III) sensors is reported in recent time. Sourav et al. reported one norbornene-derived rhodamine B, which is capable of detecting As(III) in aqueous medium up to 200 nM concentration [70]. Here, the main dominating factor is the oxidation of As(III) to As(V) in the presence of potassium iodate and concentrated HCl. During this oxidation procedure, iodine is liberated which coordinates with sensing molecule Nor-Rh, which leads to the colorimetric as well as fluorescence change. The effectiveness of this work is that the polymeric material of Nor-Rh can be used to make paper strip which will help to detect As(III) in real environmental samples. A cartoon representation is given in **Figure 9** to demonstrate the color change of polymer-coated paper strip with and without As(III).

In summary, though few reports are available for efficient detection of arsenic in aqueous medium with high sensitivity, research community continuously tries to develop sensory assay for 'on-field' application, with a tremendous impact in detection of arsenic in environmental samples with ease and real-life application.

6. Conclusion

Arsenic, having a high reactivity with cellular contents, can have diverse and deleterious effects on the cells. One of the important players of arsenic-induced toxicity is the generation of ROS, which can lead to DNA damage and lipid peroxidation. Another important effect is the arsenic exposure that causes the depletion of methyl groups in cellular milieu. Hypomethylation of promoter regions can lead to overexpression of genes which play a key role in cell proliferation, differentiation and apoptosis. As mentioned earlier, DNA hypomethylation upregulates receptors like ER-α making the cells more sensitive towards endogenous steroids. Arsenic is reported to activate PKC which activates MAPK pathway leading to the activation of various transcription factors like AP-1. AP-1 is considered as a crucial player in regulation of cell proliferation, differentiation and apoptosis. Arsenic effects extracellular matrix through upregulation of MMPs resulting in degradation of extracellular matrix having consequences in cellular migration, angiogenesis, proliferation and apoptosis. The biological effects of arsenic are so diverse that multiple

http://dx.doi.org/10.5772/intechopen.74758

mechanisms have been proposed for the toxicity of the arsenic. The mechanisms involved in the arsenic-induced carcinogenesis are also diverse and complicated. DJ-1 is a multifunctional protein that is activated upon cellular stress response. Most of the studies on DJ-1 protein are related to oxidative stress, although implication of its activity in ER stress response has been shown. The interaction of arsenic with sulfhydryl groups in proteins is considered one of the principal mechanisms which triggers the cellular responses. The binding of trivalent arsenicals to thiols in intracellular and cell surface proteins often results in aberrations of normal cellular processes including alteration of cell–cell communication. Cell–cell communication mediated by connexins, especially Cx43, the most commonly expressed connexin in different cell types, is also disrupted by arsenic binding to its highly conserved cysteine residues. In general, the effect of direct binding of arsenic species to enzyme activity cannot be ruled out in toxicity-related investigations, where other factors like the reactive oxygen species are often implicated. New methodologies are needed to analyze the health effects of arsenic and how people cope with the socioeconomic consequences of the disease. Arsenic toxicity being a global phenomenon constitutes a major public health issue, and therefore an intense research is warranted.

Acknowledgements

The study of mechanisms of arsenic-induced toxicity with special emphasis on arsenic-binding proteins is supported by the Indian Institute of Science Education and Research Kolkata (IISER-K), India. The success of this work has largely depended on the contribution and devotion of the PhD students, project students and many collaborators whose names are listed in the references.

Conflict of interest

The authors declare no conflict of interest pertaining to the contents of this chapter.

Author details

Afaq Hussain[1], Vineeth Andisseryparambil Raveendran[1], Soumya Kundu[1], Tapendu Samanta[2], Raja Shunmugam[2], Debnath Pal[3] and Jayasri Das Sarma[1]*

*Address all correspondence to: dassarmaj@iiserkol.ac.in

1 Department of Biological Sciences, Indian Institute of Science Education and Research, Kolkata, Mohanpur, West Bengal, India

2 Polymer Research Centre, Department of Chemical Sciences, Indian Institute of Science Education and Research, Kolkata, Mohanpur, West Bengal, India

3 Department of Computational and Data Sciences, Indian Institute of Science, Bangalore, Karnataka, India

References

[1] Jain C, Ali I. Arsenic: Occurrence, toxicity and speciation techniques. Water Research. 2000;**34**(17):4304-4312

[2] Bhattacharya P et al. Arsenic in the Environment: Biology and Chemistry. Elsevier; 2007; **379**(2):109-120

[3] Shen S et al. Arsenic binding to proteins. Chemical Reviews. 2013;**113**(10):7769-7792

[4] Cantor KP, Lubin JH. Arsenic, internal cancers, and issues in inference from studies of low-level exposures in human populations. Toxicology and Applied Pharmacology. 2007; **222**(3):252-257

[5] Celik I et al. Arsenic in drinking water and lung cancer: A systematic review. Environmental Research. 2008;**108**(1):48-55

[6] Roy P, Saha A. Metabolism and toxicity of arsenic: A human carcinogen. Current Science. 2002;**82**(1):38-45

[7] Hansen LJ, Whitehead JA, Anderson SL. Solar UV radiation enhances the toxicity of arsenic in *Ceriodaphnia dubia*. Ecotoxicology. 2002;**11**(4):279-287

[8] Chen B et al. Therapeutic and analytical applications of arsenic binding to proteins. Metallomics. 2015;**7**(1):39-55

[9] Hughes MF. Arsenic toxicity and potential mechanisms of action. Toxicology Letters. 2002; **133**(1):1-16

[10] Wakao N et al. Microbial oxidation of arsenite and occurrence of arsenite-oxidizing bacteria in acid mine water from a sulfur-pyrite mine. Geomicrobiology Journal. 1988;**6**(1):11-24

[11] Han FX et al. Assessment of global industrial-age anthropogenic arsenic contamination. Naturwissenschaften. 2003;**90**(9):395-401

[12] Macur RE et al. Bacterial populations associated with the oxidation and reduction of arsenic in an unsaturated soil. Environmental Science & Technology. 2004;**38**(1):104-111

[13] Páez-Espino D et al. Microbial responses to environmental arsenic. BioMetals. 2009;**22**(1): 117-130

[14] Ruiz-Ramos R et al. Sodium arsenite induces ROS generation, DNA oxidative damage, HO-1 and c-Myc proteins, NF-κB activation and cell proliferation in human breast cancer MCF-7 cells. Mutation Research, Genetic Toxicology and Environmental Mutagenesis. 2009;**674**(1):109-115

[15] Kitchin KT, Ahmad S. Oxidative stress as a possible mode of action for arsenic carcinogenesis. Toxicology Letters. 2003;**137**(1):3-13

[16] Shi H, Hudson LG, Liu KJ. Serial review: Redox-active metal ions, reactive oxygen species, and apoptosis. Free Radical Biology & Medicine. 2004;**37**:582-593

[17] Aposhian HV. Enzymatic methylation of arsenic species and other new approaches to arsenic toxicity. Annual Review of Pharmacology and Toxicology. 1997;**37**(1):397-419

[18] Chen H et al. Chronic inorganic arsenic exposure induces hepatic global and individual gene hypomethylation: Implications for arsenic hepatocarcinogenesis. Carcinogenesis. 2004; **25**(9):1779-1786

[19] Yager J, Leihr J. Molecular mechanisms of estrogen carcinogenesis. Annual Review of Pharmacology and Toxicology. 1996;**36**(1):203-232

[20] Waalkes MP et al. Estrogen signaling in livers of male mice with hepatocellular carcinoma induced by exposure to arsenic in utero. Journal of the National Cancer Institute. 2004; **96**(6):466-474

[21] Lantz RC, Hays AM. Role of oxidative stress in arsenic-induced toxicity. Drug Metabolism Reviews. 2006;**38**(4):791-804

[22] Atkinson JJ, Senior RM. Matrix metalloproteinase-9 in lung remodeling. American Journal of Respiratory Cell and Molecular Biology. 2003;**28**(1):12-24

[23] Josyula AB et al. Environmental arsenic exposure and sputum metalloproteinase concentrations. Environmental Research. 2006;**102**(3):283-290

[24] Olsen CE et al. Arsenic upregulates MMP-9 and inhibits wound repair in human airway epithelial cells. American Journal of Physiology - Lung Cellular and Molecular Physiology. 2008;**295**(2):L293-L302

[25] Huang C et al. Molecular mechanisms of arsenic carcinogenesis. Molecular and Cellular Biochemistry. 2004;**255**(1–2):57-66

[26] Dong Z. The molecular mechanisms of arsenic-induced cell transformation and apoptosis. Environmental Health Perspectives. 2002;**110**(Suppl. 5):757

[27] Schreck R, Albermann K, Baeuerle PA. Nuclear factor kB: An oxidative stress-responsive transcription factor of eukaryotic cells (A review). Free Radical Research Communications. 1992;**17**(4):221-237

[28] Thompson DJ. A chemical hypothesis for arsenic methylation in mammals. Chemico-Biological Interactions. 1993;**88**(2–3):89-114

[29] Stevenson KJ, Hale G, Perham RN. Inhibition of pyruvate dehydrogenase multienzyme complex from *Escherichia coli* with mono-and bifunctional arsenoxides. Biochemistry. 1978;**17**(11):2189-2192

[30] Stocken LA, Thompson R. Reactions of British anti-lewisite with arsenic and other metals in living systems. Physiological Reviews. 1949;**29**(2):168-194

[31] Ordóñez E et al. Evolution of metal(loid) binding sites in transcriptional regulators. Journal of Biological Chemistry. 2008;**283**(37):25706-25714

[32] Fomenko DE, Marino SM, Gladyshev VN. Functional diversity of cysteine residues in proteins and unique features of catalytic redox-active cysteines in thiol oxidoreductases. Molecules and Cells. 2008;**26**(3):228

[33] Thilakaraj R et al. In silico identification of putative metal binding motifs. Bioinformatics. 2006;**23**(3):267-271

[34] Delnomdedieu M et al. Complexation of arsenic species in rabbit erythrocytes. Chemical Research in Toxicology. 1994;**7**(5):621-627

[35] Jacob C et al. Sulfur and selenium: The role of oxidation state in protein structure and function. Angewandte Chemie International Edition. 2003;**42**(39):4742-4758

[36] Zhou X et al. Arsenite interacts selectively with zinc finger proteins containing C3H1 or C4 motifs. Journal of Biological Chemistry. 2011;**286**(26):22855-22863

[37] Torrance JW. The Geometry and Evolution of Catalytic Sites and Metal Binding Sites. Robinson College, Cambridge, European Bioinformatics Institute. University of Cambridge; 2008

[38] Brenna O et al. Affinity-chromatography purification of alkaline phosphatase from calf intestine. Biochemical Journal. 1975;**151**(2):291-296

[39] Zhang X et al. Study of arsenic–protein binding in serum of patients on continuous ambulatory peritoneal dialysis. Clinical Chemistry. 1998;**44**(1):141-147

[40] Lu M et al. Evidence of hemoglobin binding to arsenic as a basis for the accumulation of arsenic in rat blood. Chemical Research in Toxicology. 2004;**17**(12):1733-1742

[41] Faita F et al. Arsenic-induced genotoxicity and genetic susceptibility to arsenic-related pathologies. International Journal of Environmental Research and Public Health. 2013; **10**(4):1527-1546

[42] Ngu TT, Stillman MJ. Arsenic binding to human metallothionein. Journal of the American Chemical Society. 2006;**128**(38):12473-12483

[43] Merrifield ME, Ngu T, Stillman MJ. Arsenic binding to *Fucus vesiculosus* metallothionein. Biochemical and Biophysical Research Communications. 2004;**324**(1):127-132

[44] Yang J et al. Arsenic binding and transfer by the ArsD As(III) metallochaperone. Biochemistry. 2010;**49**(17):3658-3666

[45] De Maio A, Vega VL, Contreras JE. Gap junctions, homeostasis, and injury. Journal of Cellular Physiology. 2002;**191**(3):269-282

[46] Martinez VD et al. Arsenic exposure and the induction of human cancers. Journal of Toxicology. 2011;**2011**:431287

[47] Grek CL et al. Connexin 43, breast cancer tumor suppressor: Missed connections? Cancer Letters. 2016;**374**(1):117-126

[48] Barbieri L, Luchinat E, Banci L. Intracellular metal binding and redox behavior of human DJ-1. JBIC, Journal of Biological Inorganic Chemistry. 2017;**23**(1):1-9

[49] Björkblom B et al. Parkinson disease protein DJ-1 binds metals and protects against metal-induced cytotoxicity. Journal of Biological Chemistry. 2013;**288**(31):22809-22820

[50] Cullen WR. Is Arsenic an Aphrodisiac? The Sociochemistry of an Element. Cambridge: Royal Society of Chemistry; 2008

[51] Zhang X-W et al. Arsenic trioxide controls the fate of the PML-RARα oncoprotein by directly binding PML. Science. 2010;**328**(5975):240-243

[52] Lallemand-Breitenbach V et al. Curing APL through PML/RARA degradation by As_2O_3. Trends in Molecular Medicine. 2012;**18**(1):36-42

[53] Gromer S, Urig S, Becker K. The thioredoxin system—From science to clinic. Medicinal Research Reviews. 2004;**24**(1):40-89

[54] Lu J, Chew E-H, Holmgren A. Targeting thioredoxin reductase is a basis for cancer therapy by arsenic trioxide. Proceedings of the National Academy of Sciences. 2007;**104**(30):12288-12293

[55] Ybanez N, Cervera M, Montoro R. Determination of arsenic in dry ashed seafood products by hydride generation atomic absorption spectrometry and a critical comparative study with platform furnace Zeeman-effect atomic absorption spectrometry and inductively coupled plasma atomic emission spectrometry. Analytica Chimica Acta. 1992;**258**(1):61-71

[56] Samanta G et al. Calcutta pollution: Part V. Lead and other heavy metal contamination in a residential area from a factory producing lead-ingots and lead-alloys. Environmental Technology. 1995;**16**(3):223-231

[57] Chattopadhyay G et al. Determination of benzene, toluene and xylene in ambient air of Calcutta for three years during winter. Environmental Technology. 1997;**18**(2):211-218

[58] Shull M, Winefordner J. Determination of arsenic in environmental samples using neutron activation analysis. Analytical Chemistry. 1984;**56**:2617-2620

[59] Madrid Y, Chakraborti D, Cámara C. Evaluation of flow-injection in lead hydride generation-atomic absorption spectrometry. Microchimica Acta. 1995;**120**(1):63-72

[60] Hung DQ, Nekrassova O, Compton RG. Analytical methods for inorganic arsenic in water: A review. Talanta. 2004;**64**(2):269-277

[61] Forzani ES et al. Detection of arsenic in groundwater using a surface plasmon resonance sensor. Sensors and Actuators B: Chemical. 2007;**123**(1):82-88

[62] Kalluri JR et al. Use of gold nanoparticles in a simple colorimetric and ultrasensitive dynamic light scattering assay: Selective detection of arsenic in groundwater. Angewandte Chemie. 2009;**121**(51):9848-9851

[63] Wu Y et al. Ultrasensitive aptamer biosensor for arsenic(III) detection in aqueous solution based on surfactant-induced aggregation of gold nanoparticles. Analyst. 2012;**137**(18): 4171-4178

[64] Divsar F et al. Aptamer conjugated silver nanoparticles for the colorimetric detection of arsenic ions using response surface methodology. Analytical Methods. 2015;**7**(11):4568-4576

[65] Baglan M, Atılgan S. Selective and sensitive turn-on fluorescent sensing of arsenite based on cysteine fused tetraphenylethene with AIE characteristics in aqueous media. Chemical Communications. 2013;**49**(46):5325-5327

[66] http://www.who.int/mediacentre/factsheets/fs372/en/

[67] Lohar S et al. Selective and sensitive turn-on chemosensor for arsenite ion at the ppb level in aqueous media applicable in cell staining. Analytical Chemistry. 2014;**86**(22):11357-11361

[68] Dey B, Saha R, Mukherjee P. A luminescent-water soluble inorganic co-crystal for a selective pico-molar range arsenic(III) sensor in water medium. Chemical Communications. 2013;**49**(63):7064-7066

[69] Banerjee A et al. Visible light excitable fluorescence probe and its functionalized Merrifield polymer: Selective sensing and removal of arsenate from real samples. RSC Advances. 2014;**4**(8):3887-3892

[70] Bhattacharya S, Sarkar S, Shunmugam R. Unique norbornene polymer based "in-field" sensor for As(III). Journal of Materials Chemistry A. 2013;**1**(29):8398-8405